Face in Trouble – From Physiognomics to Facebook

POLISH STUDIES
TRANSDISCIPLINARY PERSPECTIVES

Edited by Krzysztof Zajas / Jarosław Fazan

VOLUME 21

PETER LANG
EDITION

Olga Szmidt / Katarzyna Trzeciak (eds.)

Face in Trouble – From Physiognomics to Facebook

Copy-edited by Soren Gauger

PETER LANG
EDITION

Bibliographic Information published by the Deutsche Nationalbibliothek
The Deutsche Nationalbibliothek lists this publication in the Deutsche Nationalbibliografie; detailed bibliographic data is available in the internet at http://dnb.d-nb.de.

Library of Congress Cataloging-in-Publication Data
Names: Szmidt, Olga, 1989- editor.
Title: Face in trouble - from physiognomics to Facebook / Olga Szmidt, Katarzyna Trzeciak (eds.)
Description: New York : Peter Lang, 2017. | Series: Polish studies - transdisciplinary perspectives, ISSN 2191-3293 ; Vol. 21
Identifiers: LCCN 2017014121 | ISBN 9783631676233
Subjects: LCSH: Face (Philosophy) | Physiognomy. | Face. | Popular culture. | Social media.
Classification: LCC B105.F29 F33 2017 | DDC 128/.6—dc23 LC record available at https://lccn.loc.gov/2017014121

Cover image: © DamienGeso / Fotolia.com

This publication was financially supported
by the Jagiellonian University, Cracow.

ISSN 2191-3293
ISBN 978-3-631-67623-3 (Print)
E-ISBN 978-3-653-06983-9 (E-PDF)
E-ISBN 978-3-631-71196-5 (EPUB)
E-ISBN 978-3-631-71197-2 (MOBI)
DOI 10.3726/978-3-653-06983-9

Peter Lang Edition is an Imprint of Peter Lang GmbH.

Peter Lang – Frankfurt am Main · Bern · Bruxelles · New York ·
Oxford · Warszawa · Wien

This publication has been peer reviewed.

www.peterlang.com

Table of Contents

Introduction

To Walter Benjamin's way of understanding, the face is the sign which manifests itself through its withdrawal, which introduces the possibility of presenting the materiality of its medium. As such, the medium becomes open to future selfconfigurations which are unavailable when an identity is fixed and stable. This concept reaffirms the position of the self as the other. It is opposed to the main metaphor of face, conceived of as a signifier of mere selfpresence. Therefore, the term 'face to face', a signifier of pure communication, can be seen as a false promise and should be redefined in the proliferation of material meanings.

Another tendency opposed to the idea of facial transparency stems from the work of Deleuze and Guattari. Their concept of facialization attacks Levinas's 'face as the other'. *A Thousand Plateaus* proposes a kind of departure from the universal domination of the face, its imperial position (facialisation obliges the subject to assume a face). Dismantling this oppressive face means performing defacialisation. Schizos are subjects who lose the dominant significations of the face.

The essays in this book were originally given as papers at a conference at the Jagiellonian University in 2015. Each essay aims to examine different tendencies which problematise the transparency of the face in ancient and modern culture. The purpose of this volume is to provide insight into some of the most important cultural phenomena related to the status of the face.

Face in Trouble: From Physiognomics to Facebook provides a wide-ranging and up-to-date overview of the recognitions and misrecognitions that constitute the universal concept of the face. Thus the figure of the face functions both as a concrete, figural representation of the human body-part, and in a more metaphorical sense, as a metaphor for the most important issues and categories in epistemology and ontology.

The articles collected in this volume adopt different perspectives and use various approaches. It includes philosophical essays, case studies influenced by cultural studies or comparative literature, and media studies. The article by Przemysław Tacik, entitled 'De-facing Philosophy: Negativity, the Government of Ethics and the Concept of the Ethical', is an insightful philosophical study on the modern tradition of thinking about the face. The author discusses modernity as well as the idea of negativity in the context of works by Emmanuel Levinas, Giorgio Agamben, Martin Heidegger and Slavoj Žižek. Levinas' imagery (including masks, mirrors, and ethical dilemmas) is a central issue being examined. Like most of the articles, Levinas' perspective and imagery are more a point of reference for discussion than a point of conclusion. The article by Mathieu Corteel examines contemporary semiotic systems of the face as well as a virtual social system like Facebook. The article is strongly influenced by works of Gilles Deleuze and Felix Guattari, as well as by Michel Foucault. The next article, contributed by Anna Rowińska, discusses the phenomenon of hysteria and depictions of madness in an inspiring way. The historical context of physiognomy and its consequences are examined through a feminist approach and a critical analysis of the patriarchal concept of female hysteria. Another essay deals with physiognomics and its influential ideas and practices. Ewelina Drzewiecka examines its presence in Renaissance culture. Her article analyses a work by John of Głogów, *Phisonomia hicinde ex illustris scriptoribus* (1518). This insightful interpretation gives us a glimpse into the historical state of knowledge about the human body and the idea of the face.

The article by Katarzyna Trzeciak examines sculptural art in its aesthetic and philosophical dimensions. The problem of the face is crucial to her interpretation, though it also includes general reflections and observations on aesthetics. Modern literature is the topic of the article by Adrianna Alksnin. In the article significantly entitled 'Me, Myself and My Face: Pirandello Meets Gombrowicz' she explores the issue of the disintegration of the modern self. The comparison of the two novels also leads the author to investigate the relations between face and identity.

The following chapters of *Face in Trouble* contain articles focused on film, television, social media and games. The first one is written by Beja Margitházi. By exploring the concept of the Uncanny Valley, the author examines technology used for building characters, especially computer-generated ones, in cinema. Another case study pertains to a fragment of the *Homeland* TV series. This kind of spectacle of the face in contemporary television is at the core of Olga Szmidt's article. A suicide tape also is interpreted as an act of confession. Apart from interpreting this specific case, the article focuses on contemporary approaches to the idea of authenticity.

While the article by Magda Ciereszko is also focused on contemporaneity, it is of different complexion. Her interpretation of Instagram and other social media images of the face and body gives the reader a chance to explore the latest phenomena. The face and the body are not only ideas in these cases. They are rather well-designed practices, images as well as products. The article which concludes the volume is focused on modern role-playing games. The voice of the avatar and the face of the other are used to play with ideas of agency and control. The essay by Mateusz Felczak leads us to surprisingly new ways of considering the face and its usage in contemporary art and entertainment. The problem of the player's presence and agency are investigated in the context of twenty-first-century role-playing games.

All the articles discuss the idea and representations of the face in different ways. We begin *Face in Trouble* with philosophical perspectives and historically influential physiognomics, and we end with role-playing games and faces on Instagram. One might say it is hard to open a discussion about a phenomenon of such complexity from such diverse perspectives and fields of knowledge. In our view, the idea of the face requires this diversity. Its beauty and ugliness, complexity and naivety, its weakness and power fascinate and disturb at the same time. We shall face this phenomenon and its representations in different periods and cultures. Although, as with gender, one thing is immutable: the power of the face and all the trouble it causes.

Olga Szmidt
Katarzyna Trzeciak

About the Contributors

Adrianna Alksnin is a PhD Candidate at the Faculty of Polish Studies (Center for Anthropology of Literature and Culture Studies), Jagiellonian University in Krakow. She has published in *Teksty Drugie*, *Przegląd Humanistyczny* and *Pamiętnik Literacki*. She is also a student of the Faculty of Drama Directing at the National Academy of Theatre Arts in Krakow.

Magda Ciereszko is a PhD Candidate at the Institute of Applied Polish Studies, Department of Rhetoric and Media of the University of Warsaw. She graduated from the Faculty of Polish Studies and the Faculty of Journalism and Political Science at the Univeristy of Warsaw. Her Master's thesis examined relations between pornography and mainstream culture. Her PhD thesis focuses on the concept of obscenity in modern advertising. She also lectures on the poetics of advertising, critique of advertising, social media and copywriting. She has been working in advertising since 2012. Selected publications: 'Performing for the followers – social media as a platform for the artist', *Contemporary Lynx*, Vol. 2, pp. 68–71 (2016); 'Popularna pornografia? O rodzajach pornografii funkcjonujących na ziemiach polskich na początku XX wieku i jej relacji z ówczesną kulturą popularną', *Kultura i historia*, Vol. 29 (2016).

Mathieu Corteel is a lecturer at the Institute of Psychology of Paris-Descartes University (Paris V), a seminar manager at Collège International de Philosophie (CIPH) and a PhD Candidate at Paris-Sorbonne University (Paris IV). He develops an epistemological history of the numerical method in medical sciences from a Foucaldian perspective. His research especially focuses on the emergence of big data. His most important article is 'La mort *in vivo* et le vivant *postmortem*', *Revue Le Philosophoire*, 1/2016 (No.45).

Ewelina Drzewiecka holds degrees in Classical Studies (BA, 2007; MA, 2009) and in History of Art (MA, 2011), both from the Adam Mickiewicz University in Poznań, Poland. In 2012, she became a PhD Candidate in Neo-Latin Literature at the Jagiellonian University in Krakow. She also studied at the University of Perugia (2010/2011) and the University of Bologna (2015/2016; 2017), participated in numerous conferences and has published a number of research papers. She has been awarded national and international grants (Italian Government Scholarship, Lanckoroński Foundation Scholarship) and has run several research projects in

Poland, Italy and the United Kingdom. Her research focuses on Neo-Latin Studies, Early Modern Literature and the fine arts.

Mateusz Felczak is a PhD candidate at the Institute of Audiovisual Arts, Jagiellonian University in Krakow. His research interests include cognitive capitalism, e-sports, Fluxus, object-oriented philosophy and the mechanisms for modeling the reception and distribution of computer games, especially in the AAA segment. He is a member of the *ETC* research/artistic group, editor of the video games section of *EKRANy* magazine, and collaborates with the *Glissando* magazine. He has published in *Kultura Popularna, Kultura Współczesna, Homo Ludens, Praktyka Teoretyczna* and *Ha!art*.

Beja Margitházi is an assistant professor at the Film Studies Department of the Eötvös Loránd University (Budapest, Hungary). Her research interests include history and theories of film and visual culture, contemporary film (analog-digital nostalgia, sensual aspects, multi-sensory experience) and post-communist Hungarian cinema. She is the author of the book *Az arc mozija. Közelkép és filmstílus* [The Cinema of the Face: The Close-up and Film Style] (2008) and co-editor of the anthology *Vizuális Kommunikáció* [Visual Communication] (with Ágnes Blaskó, 2010). Her studies, critical essays and translations have been published in various Hungarian and English anthologies, periodicals and magazines. She has been a co-founder and editor of Transylvanian film magazine www.filmtett.ro.

Anna Rowińska graduated in cultural studies and art history from the University of Warsaw. She is a curator of contemporary art exhibitions. She works at the Polish Modern Art Foundation in Warsaw, is a curator of the Keret House Residency Program and a coordinator for the Biuro Wystaw contemporary art gallery.

Olga Szmidt is currently a PhD Candidate in Literary Studies (expected completion: 2017) at the Faculty of Polish Studies, the Jagiellonian University in Krakow. She is preparing a dissertation on authenticity in twenty-first century culture. She graduated in Polish Philology and Literary Criticism. She has published a monograph on the epistolography of Stanisław Ignacy Witkiewicz (*Korespondent Witkacy*, Universitas 2014), which received the J. J. Lipski Award for the best Master's thesis in the humanities and social sciences. Her essays and articles have been published in *Teksty Drugie, Pamiętnik Literacki, FA-art* and in Polish and international monographs. Her professional interests include authenticity, contemporary culture and the modern self. She is also a non-fiction writer and a literary and cultural critic.

Przemyslaw Tacik is an assistant professor at The Institute of European Studies, the Jagiellonian University in Krakow. He holds PhDs in philosophy and law. His interests include contemporary continental philosophy, Jewish philosophy and political philosophy, as well as international law and critical legal studies. Apart from papers in English and in Polish, as well as translations from French, he has authored two monographs (in Polish): *The Sociology of Zygmunt Bauman* (2012) and *The Liberty of Lights: Edmond Jabès and the Jewish Philosophy of Modernity* (2015; an English translation is scheduled for publication in 2019).

Katarzyna Trzeciak is a PhD Candidate at the Faculty of Polish Studies, the Jagiellonian University in Krakow. Her thesis explores the metaphor of sculpture in the literature of early twentieth century. Her research interests focus on phantasmatic criticism and psychoanalytic studies in literature.

Przemysław Tacik

De-facing philosophy: Negativity, the government of ethics and the concept of the ethical

This paper explores the metaphor of the face in philosophy, principally in the works of Emmanuel Levinas. The face plays a structuring role in his ethics: it organises the discharge of negativity that is subsequently used to found an ethical injunction. Yet such a usage of face-clad negativity – inherited, through Heidegger, from modern Western philosophy – demands profound reconsideration. I argue that modern thinking often resorts to what may be called 'face philosophy' – a special method of arranging negative energy for seemingly external (e.g. ethical) purposes. Such a re-organisation of negativity, enticing as it may seem, cannot be unproblematic in confronting the Heideggerian legacy. We need to rethink *the government of ethics*, the methods of arranging negative powers to dispense ethical injunctions. To this end, the paper forwards an analysis in terms of *the ethical* – a concept which, as a counterbalance to Schmitt's 'political', allows us to question each step within ethics as the potential work of negativity.

Keywords: face, ethics, Levinas, Heidegger, Schmitt

> It knew no lapse, nor Diminution –
> But large—serene—
> Burned on—until through Dissolution –
> It failed from Men—
>
> I could not deem these Planetary forces
> Annulled—
> But suffered an Exchange of Territory—
> Or World—[1]
>
> *Emily Dickinson*

Philosophy rarely seems to deal with the face, either as a literal object or as a metaphor; had it not been for Emmanuel Levinas, the category of the face would probably never have gained a philosophical existence unto itself. Naturally, in the 'mobile army of metaphors' (to use Nietzsche's famed expression [Nietzsche 1999, p. 145]) which philosophy deploys in order to construct its discourse, the face has always been among the lesser-known foot soldiers. It has always been

1 Emily Dickinson, *The Complete Poems of Emily Dickinson*, ed. by Thomas H. Johnson, Back Bay Books, New York-Boston-London, p. 272.

present, but more as a handy colloquial tool which can be readily adapted as a philosophical crutch. Levinas' thought should be credited for breathing new life into this hitherto marginal category.

Nevertheless, while the face has a distinct and quite precise meaning in Levinasian thinking, it has much greater philosophical potential. Even if the face itself is not frequently an explicit concept, it might prove useful to *unmask* some philosophical stratagems and narratives which are, in fact, structurally built around the face. What I will be seeking in this paper is not the face as a distinct philosophical category or a metaphor of philosophical discourse, but as *a trace of structure* (re)organising the course of the philosophical narrative. As such, its usage can be re-constructed as a fairly common gesture in philosophy which, as we shall see, is particularly significant in stratagems of contemporary ethical thought. With all due caution, it might even be claimed that there is a 'facial paradigm' in philosophy, a particular way of curving the argument that structures itself in a way best grasped by the motif of the face.

Following a Derridean train of thought, we might suggest that once this facial paradigm of philosophy is outlined, it should not be criticised—this would be neither possible nor useful. What should be done, perhaps, is to de-face philosophy: to overwrite its bare façade with marks that obstruct transmission and representation between what is covered by the face and the face itself. In other words, de-facing would be equivalent to destabilizing the predetermined flow between positions set by particular forms of philosophical argumentation.

This de-facement will provide a starting point for reconsidering the role and possibility of ethics in and after Levinas. The horizon of reasoning in this paper is outlined by the concept of the ethical, which, as I will attempt to demonstrate, might supersede ethics.

From the Levinasian figure of the face to the facial paradigm in philosophy

The figure of the face is instrumental in Levinas' philosophy. It is a powerful source of inspiration for reflecting on the direct relation between two people and between a human being and God. Moreover, it might be rightly dubbed the emblem of Levinasian ethics which introduces the crux of the ethical injunction in a readily accessible manner. Considering that a surprising number of studies have been devoted to this ethics, there is little point in reconstructing it here, given also that it is beyond the scope of this paper. Let us then summarise briefly—with intentional simplification—the exemplary use of this figure in Levinas' discourse.

The following fragment comes from *Conversations with Emmanuel Levinas* by Michaël de Saint-Cheron:

> [...] isn't man's relation to man already in itself what is essential in man's relation to the Eternal? This is not at all a verbal reduction of one to the other. The discovery of the other man's face in the responsibility toward him is how we hear God's voice. [...] God's coming to mind is certainly contemporaneous with the responsibility assumed toward the other's face.
> [...] I want to insist on the fact that seeing the face is not like perceiving something. It's very important to grasp this. It's not like seeing a thing, seeing a painting, not at all. It's an immediate relation of responsibility and consequently the word of God. Indeed, there's quite a difference here with Aristotelian logic. There are three of us when there are two of us. What I'm saying shouldn't be understood in terms of substance but in terms of classical philosophy. This third place is unique. God is present. We're already responding to him. [de Saint-Cheron 2010, pp. 33–34]

Although this fragment pertains to Levinas' late philosophy—which is more marked with explicit religious references than his earlier thought—it exemplifies the most common Levinasian usage of the figure of the face. The face has an attribute of centrality or concentration: it channels and positions the relation between what it covers and 'the front' of philosophy, namely the gist of the ethical message. A man finds a man (the gender aspect of Levinas' discourse will be mentioned later) through his face, which automatically establishes a relationship of responsibility and summons God as the third partner in a dialogue—a partner who is, simultaneously, its guarantor.[2] The face reveals responsibility as something which predates the relation. In fact, it turns the one to whom it is revealed into a condition of primordial indebtedness. Levinas underlines the directedness of the relation sustained by the face. This relation is overwhelming, unquestionable and, for all its restrictions, transcendent. Due to its bifurcated nature (the relation to another human being and to God), it is both stable and unstable, present and absent; the face, however, acts as its supporting agent.[3]

2 This concise presentation for the purposes of this paper must necessarily omit all the differences between the particular stages in Levinas' thought. Probably the most important of these concern the displacement of the role of total responsibility in relation to subjectivity in *Totality and Infinity* [Levinas 1991b] and in *Autrement qu'être ou au-delà de l'essence* [Levinas 1974]. Despite this evolution, the pivotal role of the figure of the face is preserved, so for the purposes of this paper it need not be elaborated upon.

3 A more developed analysis of the face as the emblem of responsibility might be found in Levinas' first famous work, *Totalité et infini*, especially in Part III. Cf. also Levinas 1990, p. 8f.

The Levinasian use of the figure of face might be criticised in many regards. Although listing them is not the aim of this paper, it should be briefly mentioned that at least three major types of critique might shed light on the seemingly innocent ethical reflection concentrated in the figure of face. The first one is the most intuitive: The Levinasian ethics of ultimate and unconditional demand is as radical as it is ineffective. The responsibility one has towards the other is total, which practically paralyses the execution of the moral injunction. Although Levinas did a great deal to supplement his philosophy with a social dimension (especially after Derrida's critique in *Violence and Metaphysics* [Derrida 2001, pp. 97–192], which pointed out the narrowness of present face-to-face relation), the purified (and isolated) face-to-face relation occupies the core of his ethical imagination. The presence of God as the third partner within this relation hardly opens up its somewhat claustrophobic closure, to say nothing of alleviating the unconditionality of the call for responsibility. As a result, the cornerstone of Levinasian ethics appeals to both the philosophical and the common imagination, but practically throws them in a kind of self-righteous day-dreaming; the injunction is too demanding to be effectively enacted, so it is reduced to an object of contemplation. This critique pertains to intuitions both of more pragmatically oriented ethics and of purely speculative ethical thought, for which the paralysis of the face-to-face relation engenders excessively dogmatic consequences.[4]

The second standard critique of how Levinas uses the figure of face concentrates on the notable speciesism and patriarchalism of his argument. Used as an imaginary core of ethical reflection, the face is fundamentally confined to the human face. With the primacy of total responsibility for which we are called in the epiphany of the face, all responsibility towards non-human animals, not to mention other beings, is relegated to the back burner, if not obliterated. In this regard, Levinasian ethics inherits the pro-human bias of what it struggles to overcome: early Heideggerian thought.[5] The patriarchalism, in turn, is not just in

4 In this regard the fate of Levinasian inspiration is particularly telling in terms of social thought. Zygmunt Bauman, who drew abundantly upon Levinas' philosophy in his reflections on ethics in postmodern society, had to struggle with the paradigmatically a-social character of this ethics. Although Levinasian ethics contributed to Bauman's overall approach—attempting to single out the ethical situation from socially imposed morals—there remained a vast rift between total responsibility for the Other, experienced in the (utterly private) epiphany of face, and ethics on a social scale. In order to find a concept to bridge the two, Bauman sought to reinterpret Levinas' category of justice, but with limited success. Cf. Bauman 2001, pp. 181–183.

5 Heidegger's philosophy from the *Sein und Zeit* period privileges *Dasein* as the site of the proper understanding of being (*Sein*). For all its ambition to rework how the

the lack of linguistic gender sensitivity here (the French *homme* is hardly a more alluring universal term for 'human being' than the English *man*), but also in the ostensible focus on the symbolic relation between men as the universal core of ethics. Surely, the Levinasian face is not simply the face of a man, but this figure contains the false universality of a man-to-man relationship.

As a result, the ethics built upon this metaphor emerges as a pact between male human beings guaranteed by God. Total responsibility, the focal point of Levinas' reflections, is destabilised by of women, because their faces—at least according to *Totality and Infinity* [Levinas 1991c, pp. 294–295]—are marked by the parasitic attributes of beauty and eroticism (implicitly: to the male eye). Apart from Levinas' clearly patriarchal remarks on women, which sometimes border on glaring insults, especially in his early writings,[6] the primacy and unconditionality of the ethical

---

human being is portrayed in modern thought, *Dasein* is clearly confined to the *Dasein* of human beings. The neat separation between human and non-human animals is the cornerstone of Heidegger's anthropology in this period, most notably expounded in the seminar *Die Grundbegriffe der Metaphysik* from 1929–1930. The key sentence of one part of this seminar distinguishes between material objects, which are worldless (*weltlos*), animals, which are world-poor (*weltarm*), and human beings, who are world-forming (*weltbildend*), and therefore essentially distinct from non-human animals [Heidegger 1983, p. 263]. This distinction owes some of its sharpness to the phenomenological sources of Heidegger's thought (and, by extension, to Cartesianism), as well as to the anthropology of Jakob von Uexküll.

6 An early text by Levinas—*Le temps et l'autre* (adapted from his lectures delivered in 1946–1947)—provides a concise, yet telling example of how his musings on responsibility, which aspire to universality and ethical sublimity, are haunted by the violence of patriarchalism. In the fragment entitled 'Eros', Levinas attempts to link the otherness of the Other with the difference between the sexes, claiming that it is in the opposition between men and women that otherness is at its purest [Levinas 1991a, p. 77]. This opposition, however, is not defined objectively, but as the difference which is embodied by women; therefore it is assumed that the philosophising subject is necessarily a man, speaking to men, referring to an implicit community of men, for whom women appear as 'absolutely different' [*ibid.*]. The specificity of this difference lies in the mystery of woman, who withdraws, evades, hides, and covers herself through modesty [*ibid.*, pp. 79–80]. Conflating the woman with the Other, Levinas smoothly proceeds to claim that 'the fact of being resistant to our power is not a greater strength than ours' [*ibid.*]. The word 'ours' connotes the seemingly universal community of gazing men, whose object transcends and eludes the subject-object relationship, thereby bringing erotic interest and total indebtedness to a point of indifferentiation. This peculiar mix could hardly be more repellent in its demonstration the true dimension of fetishism of the Other in Levinas' thinking.

situation privileges the man-to-man relationship, purportedly the purest of all (since it is allegedly not tainted by eroticism). In fact, there are no other relationships than this (e.g. woman-to-woman) except for the man-to-woman relationship, in which the woman is not the 'pure' Other, but either the-feminine-Other constructed on the basis of the (simple) Other with an admixture of sexist visions of the total otherness of 'the other sex', or in which the woman incorporates a supplementary (in the Derridean sense) attribute of eroticism. Consequently, in both forms, the cornerstone of the philosopher's ethics is a trap of self-refining masculinity which is closed to other ethical sensibilities.

Turning to Levinas' speciesism, we find it is again rooted in the phenomenological (and early Heideggerian) primacy of consciousness,[7] imagined as exclusively human. Levinas' God is a God of human beings and, perhaps, of conscious beings; as a result, total responsibility is clearly their domain. Even though this ethics presents itself as humble in spirit, the primacy of human beings is undeniable.

Thirdly, the Levinasian use of the figure of face might be criticised as closed to the properly traumatic dimension of the face. To do the philosopher justice, it should be noted that the encounter which lays the foundations of his ethics is far from a sentimental meeting of two people well-disposed towards each other, as popular portrayals seem to claim. Indeed, the encounter is traumatic, as it breaks into the lifeworld with all the ruthlessness of an ethical injunction. Nevertheless, Levinas seems to omit the profound analysis of all quintessentially traumatic face-to-face relations—such as one founded on true sadism (in which the suffering face of the other acts as the target of sadistic action), or the face in Agamben's examples of bare life (e.g. in the face of a *muselman*, a prisoner of death camps whose appearance oscillates between humanity and a real, almost mechanistic life process [cf. Agamben 1998]). This more or less explicit critique, deducible from Agamben's[8] and Žižek's [2006, pp. 145–190] [9] writings, questions the possibility

7 This approach is irrefutably biased by presentism, that is, a focus on present, face-to-face relations; this feature provided Derrida with a wealth of material for developing the foundations of deconstruction in *Writing and Difference*.

8 Incidentally, Agamben presented eyes not as a figure of transparent expression, but of inert resistance ('*per la loro ostinata resistenza all'espressione, per il loro torpido*'); as such, the human face becomes an embodiment of 'abyssal opacity' ('*abissale opacità*') and 'misery' [Agamben 2002, pp. 113–114].

9 Parenthetically, Levinas' discourse of the radical expropriation of the subject in the epiphany of face—and of the flaw in the very fact of existing (which necessarily entails violence)—was intriguingly critiqued by Žižek, who pointed out that the true danger lies not in existence as such (which might be justified by the Spinozian concept of

of ethics unambiguously based on the archetypal face-to-face encounter. It might as well be a source of utter terror and repulsion which undermines ethics.

There might be, however, another kind of critique which brings us to challenge philosophy's whole 'paradigm of the face'. This critique underlines the link between ethics and radical destitution—a link which should be the ultimate safeguard of the ethical injunction in Levinas' philosophy. Turning ethics into the new *prote philosophia* draws on the energy of Heideggerian thinking but attempts to change its vector to serve ethics. Nevertheless, the philosophical imagination stands by the primacy of a radical, purified situation which—in all its bareness—should underpin unconditional responsibility in all other situations. *This drive toward purity and origin, however promising it might appear in ethics, is characteristic of philosophy's 'paradigm of the face'.*

In more general terms, however, is there such a thing as a 'philosophy of the face'? If we assumed its existence, it could be a kind of philosophical narrative which focuses on the face, taken either as a metonymy or a metaphor. It might also be a narrative that draws from the face, using it as a less important figure of speech—subtle, unassuming, yet decisive for the self-positing of the discourse. A 'philosophy of the face' might involve a positive appraisal of appearance or superficiality.

Yet in this paper, the philosophy of the face will be understood as a more profound, though elusive type of philosophical discourse. It will be a philosophy of inner imbalance, of distinct inner appraisal of key elements (for instance, face and matter). It will be a kind of thought that uses an ethically and politically charged central concept; it serves as the face of this philosophy, but its existence and very way of functioning hinges on the hidden presence (or, should we rather say, present absence) of a much darker area, of crude philosophical matter veiled by the face-concept. As such, it is a philosophy which first organises points within itself and the flow of narrative energy between them, only to use it and sap the accumulated energy for the purposes of the argument.

It would not be quite adequate to call such a philosophy binary, as it falls short of the proper Derridean understanding of the term. It might elude the juxtaposition of its two 'parts' as an explicit, let alone stable opposition. Rather, in a quasi-Deleuzian fashion, it organises its inner flow between two distinct areas; it

---

conatus, granting me the very same right to be as it does to the Other), but in human existence as the center of negativity, as a 'black hole' among beings [Žižek 2014, pp. 328–329].

permanently re-configures the frontier of 'face' and 'matter', disguising its inner imbalance.

The crude matter under the face

If such a concept of a 'philosophy of the face' is assumed, it might be examined only through positional analysis, which captures dynamically changing positions of each element in its protean relation to other elements. In fact, this is the self-establishing and self-organising division of many modern philosophies which conceal their true other within themselves. Perhaps any philosophy which claims to confront the Other or simply takes him/her for a figure of discourse is organised around its own internal split, in which a 'philosophical other' provides the energy necessary for the whole project.

Having formulated such assumptions, let us ask: what is this primordial matter that the philosophy of the face conceals? Once again Emmanuel Levinas provides us with a useful example. I am going to draw upon an article of his, *Le Dit et le Dire*; even though this text was recycled in *Otherwise than Being*, taken by itself—in its rawer form—it sheds more light on the mode of argumentation in the philosophy of the face. It is a late text by Levinas, attempting to overcome the simplicity of the face-to-face encounter. Nevertheless, the figure of the face plays a key role here.

As we have already noted, the epiphany of the face plays a key role in his ethical imagery and reasoning. Confronted by the face of the Other, I am called upon to find myself responsible for him (parenthetically, Levinas uses the pronoun *il*, 'he/him,' in place of *l'Autre*, 'the Other,' which is another sign of the selectivity of his ethical imagination). In his mature thought, however, the nitty-gritty of Levinasian ethics is supplemented with the opposition between *le Dit* ('the said') and *le Dire* ('the saying'). As portrayed in the above-mentioned article, *le Dit* is the dimension of the existing language which has ossified into a device producing names and beings. Its substantified nature fits into the structure of discourse:

> Le langage comme Dit peut donc se concevoir comme un système de noms identifiant des entités et, dès lors, comme un système de signes doublant les étants désignant des substances, des événements et des relations par des substantifs ou par d'autres parties du discours dérivées des substantifs, désignant des identités—bref, *désignant*. [Levinas 1971, p. 34][10]

10 'Language as the Said can therefore be conceived of as a system of names identifying entities and, hereafter, as a system of signs doubling beings that denote substances, events, and relations with nouns or other parts of speech derived from nouns, denoting identities—in short, *denoting*.'

The dimension of the *Dit* is based on the close correlation between beings and names, joined by *designation*. It is focused on the institution of the name, which—as language's anchor in the world of beings—guarantees an iterable reference to objects and provides a solid framework of time:

> L'étant qui apparaît identique dans la lumière du temps est un Dit. [Levinas 1971, p. 29][11]

The *Dire*, on the contrary, names the dimension of the becoming flow of language. It is creative, determining, but not-yet-determined, just as verbs are:

> Le langage ne se réduit pas à un système des signes doublant les êtres et les relations—conception qui s'imposerait si le mot était Nom. Le langage serait plutôt excroissance du verbe. Et c'est déjà en tant que verbe qu'il porterait la vie sensible—temporalisation et essence de l'être. La sensation vécue—être et temps—s'entend déjà dans le verbe. [Levinas 1971, p. 27][12]

The *Dire* is not focused on beings/names (as the *Dit*), but on the temporal and ontological fluidity verbs produce. It opens the solidness of names, breathing life—and time—into them. Both dimensions are dialectically intermingled:

> Dire corrélatif du Dit, identifiant un rassemblé dans un *présent*, se le re-présentant et l'arrachant ainsi à la labilité du temps. Le Dire tendu vers le Dit et s'absorbant en lui, le Dire *corrélatif* d'un Dit, nomme un substantif identifié dans la modification du temps vécu qui éclaire (et qui peut elle-même s'identifier dans un dit). Désignation identifiante qui ne vient pas s'ajouter du dehors au phénomène par l'effet d'un code conventionnel réglamentant l'usage d'un système de signes. C'est par le déjà dit que les mots, en tant que mots de tel ou tel autre vocabulaire historiquement constitué, trouveront une fonction de signe et un emploi. [Levinas 1971, p. 29][13]

---

11 'A being which appears identical in the light of time is a Said'.

12 'Language cannot be reduced to a system of signs doubling beings and relations—a concept which would impose itself if the word were a Name. Language would rather be the excrescence of the verb. And it is as a verb that it would carry sensible life—temporalisation and the essence of being. The experienced sensation—being and time—already resounds in the verb.'

13 'The Saying is correlative of the Said, identifying a collective in the *present*, re-presenting it to itself and thus wrenching it from the lability of time. The Saying outstretched towards the Said and piercing it, the Saying *correlative* to a Said, names a noun identified in the modification of the experienced time which it illuminates (and which can identify itself in a Said). Identifying denotation which just does not add itself to the phenomenon from the outside through a conventional code controlling the usage of a system of signs. It is by the already said that words, as words of this or another historically constituted vocabulary, will function as signs and be applied.'

The opposition of the *Dire* and the *Dit* determines the dynamics of language in its ossification and dissolution, in the aporetic split between the past, preserved in the active memory of names, and the future, which is opened up by verbs.

This opposition ostensibly aims to criticise ontology. Regardless of the somewhat clichéd background of this distinction, with its echoes of Romantic oppositions, it brings Levinas' philosophy closer to the dynamics of the late Heidegger's *Er-eignis*, although in a different configuration. The perception of verbs—often coined from nouns—as factors of de-essentialisation is clearly indebted to Heidegger's use of language, but the opposition between the *Dire* and the *Dit* remains an original contribution of Levinasian thought. What immediately attracts our attention is the undisguised attempt to turn the apparently obvious opposition into a more profound and ambiguous philosophical device of notions which are inextricably intertwined. Levinas is far from setting the ossified past against the openness and fluidity of the present; his argument—partly due to the convolution inherent in the philosopher's discourse—carefully avoids fitting various oppositions neatly into the same pattern of difference between the *Dire* and the *Dit*. Since it is not the aim of this paper to elucidate Levinas' thought, it suffices to give one example of exquisite reasoning which does not juxtapose the *Dire* and the *Dit*, but attempts to present them as dialectically related moments in the self-engendered diachrony of time (apparently drawing upon Kant's musings on time as a form of sensibility [Kant 1910, pp. 71–89]:

> La signification du Dire va au-delà du Dit : ce n'est pas l'ontologie qui suscite le sujet parlant, c'est la signifiance du Dire au-delà de l'essence qui justifie l'exposition de l'être ou l'ontologie. Car le laps de temps, c'est aussi de l'irrécupérable, du réfractaire à la simultanéité du présent, de l'irréprésentable, de l'immémorial, du pré-historique. Avant les synthèses d'appréhension et de reconnaissance—s'accomplit la synthèse absolument passive du viellissement. C'est par là que le temps *se* passe. L'immémorial n'est pas l'effet d'une faiblesse de mémoire, d'une incapacité de parcourir les grands intervalles du temps, de ressusciter de trop profond passés. C'est l'impossiblité pour la dispersion du temps de se rassembler au présent, la diachronie insurmontable du temps, un au-delà du Dit. C'est la diachronie qui détermine l'immémorial, ce n'est pas une faiblesse de la mémoire qui constitue la diachronie. [Levinas 1971, pp. 30–31][14]

---

14 'The meaning of the Saying goes beyond the Said: it is not ontology that creates the talking subject, it is the meaningfulness of the Saying beyond essence that justifies the exposition of being or ontology. Because the lapse of time is also somehow the irretrievable, the refractory to the simultaneity of the present, the unrepresentable, the immemorial, the prehistorical. Before the syntheses of apprehension and recognition the absolutely passive synthesis of aging completes itself. It is through this that time passes *itself*. The immemorial is not an effect of a weakness of memory, of inability

However, towards the end of the above-cited article, the subtlety of the dialectical argument seems to run aground; *this is precisely the moment in which thought seeks to discharge the accumulated potential with the figure of the face.*

After analysing the dialectics between the *Dire* and the *Dit,* Levinas proceeds to link this opposition to his idea of total responsibility. This step instantaneously disrupts the momentary, self-reestablishing balance between the two concepts. The *Dire* is granted primacy, as the dimension which allows one to respond to the Other and to take responsibility for *him*. As soon as the proper *point de capiton*, to use a Lacanian term, is posited, the reasoning veers towards a less ambiguous juxtaposition of the *Dire* and the *Dit*. It creates an impression of nothing less than a political intrusion, through which some external, predetermined forces re-organise the hitherto autonomous argument. All the possible, but previously ambiguous virtues of the *Dire* as compared with the *Dit* now become fully pronounced.

The destabilising power of the *Dire* emerging at the crossroads between language and time is now diverted to fuel the ethical device of total responsibility sealed by the figure of the face. The intervention of the *Dire* is much more direct than the effect it has had on the *Dit*; it allows the self-centred closeness of intentionality to be pierced [Levinas 1971, p. 42]. This 'piercing' means that the subject no longer uses language to convey messages, it becomes a sign itself: *its position is wholly displaced, so that it finds itself in a situation of total destitution*. Levinas remarks:

> Le sujet du Dire ne donne pas signe, il se fait signe. Il est-pour-un-autre, non point en recueillant de cette signifiance entendue comme acte, une *essence*, une condition dans l'être, une fonction ou un rôle, il est pour un autre en signifiant à l'autre cette signifiance même, précisément *en parlant*. [Levinas 1971, p. 42][15]

Therefore the destabilising power of the *Dire* contributes to abolishing the subject as the subject of speech. Instead of being mediated to the Other by a sign, the subject lays itself totally open to the Other through adopting a position of coincidence with the sign. Since the *Dire* took over, the *Dit* unambiguously stands for

---

to cross big intervals of time, to revive too deep a past. It is the impossibility of the dispersion of time to aggregate itself in the present, the insurmountable diachrony of time, a beyond of the Said. It is diachrony that determines the immemorial, it is not a weakness of memory that constitutes diachrony'.

15 'The subject of the Saying does not give a sign, it makes itself a sign. He is-for-the-other, not at all obtaining this meaningfulness construed as an act, an *essence*, a status of being, a function or a role, he is for an other meaning for the other this very meaningfulness, precisely *in speaking*'.

unambiguous vices—sovereign, subjective demands which must be overcome in the act of assuming total responsibility for the Other.

The *Dire* concentrates the power of radical refining, which purifies the subject of its wants and particularities. Levinas' discourse is positively uncompromising; he resorts to words which in contexts other than ethical injunctions would be associated either with high Modernism or, after World War II, with the utmost degradation of the human being. The philosopher describes the self-emptying, expropriation [Levinas 1987, p. 26] and self-immolation of the subject for the sake of a responsibility for the Other. Ultimately, before the radicality of the ethical injunction, there is almost nothing left except for the core of the abolished subject (which, through the process of '*dépouillement jusqu'au noyau, ab-solution*', turns into 'un *sans être,* un *sans complexion,* un *inqualifiable quelqu'un*' [Levinas 1971, p. 43]). Although Levinas attempts to shift the vector from the negative philosophical energy with its more-or-less obscure links to the total destitution of humanity, his thought still draws energy from the reservoir of negativity. This trick allows him to seek hope for ethics not in times when the social fabric is dense, and moral behaviour is enforced through a complex network of norms safeguarded by strong social hierarchies, but in times of alienation, anomy, meaninglessness and dehumanisation [Levinas 1971, p. 48].

Having outlined the development of Levinas' argument in *Le Dire et le Dit*, we can now proceed to trace the strategies of 'the philosophy of the face' in this article.

We have already assumed that the philosophy of the face might be intuitively approached as a philosophy which groups its inner forces with a powerful ethically and/or politically charged concept, releasing a predetermined difference of potentials and channelling it in order to produce an injunction. If we apply this tentative notion to *Le Dire et le Dit*, two main observations might be made.

Firstly, Levinas' argument is clearly split into two parts: the main part of the article uses a carefully balanced argument which deftly links the *Dire* and the *Dit,* dialectically and dialogically. The former seems to be granted a kind of primacy, as the general tone of the narrative is slanted towards the flexibility and indetermination of happening. Nevertheless, a discharge lurks on the horizon. The apparent equality between the two nearly balanced concepts has an underlying tension, as if the relation between the *Dire* and the *Dit* cannot be contained within it. The *Dire* starts to gather and structure the energy of negativity,[16] which, in all its dynamism, breaks solid boundaries and ossified identities. The *Dit* begins to connote negatively valourised structures: sediments of language, essences, closure.

---

16 The link between negativity and the imagery of energy *behind* thinking was grasped by Heidegger with reference to Hegel [Heidegger 1993, p. 27].

As soon as the ground is sufficiently prepared, the key concept of responsibility emerges like lightning from the clouds. With this sudden discharge, the *Dire* and the *Dit* turns into an opposition whose parts are unambiguously valourised. With regard to our tentative notion of 'the philosophy of the face', it might be claimed that *in Levinas' discourse the concept of responsibility acts as the figure of the face*, which, fortunately, is explicitly taken for an emblem of this thought. The face, construed as a figure of philosophical discourse, connoting a particular concept, emerges within it in order to polarise, redirect and channel the potential which has accumulated prior to its entrance. It provides a discharge for the tension mounting in the course of the argument. As such, it acts retroactively, re-defining the sense of the play between particular concepts and the flow of energy between them. Even though the dialectical richness that was elaborated before its appearance is not explicitly disavowed, it is furnished with a conclusion which effectively simplifies the network of dialectical links. Nevertheless, the subtlety of the foregoing dialectical reasoning contributes to the solidity of the following argument, *whose ongoing simplification does not give the impression of a predetermined, assumed device, but seems to flow naturally from the inner dynamics of the previous reasoning.*

In this regard, the figure of the face plays a crucial role in philosophical discourse. It governs the displacement of positions around which the argument flows. Linking the apparent richness of profoundly problematised, dialectical reasoning with radical clear-cut oppositions, *it creates and organises the process of simplification*, which is a unilateral type of philosophical argument aiming to produce, govern and discharge the potential gathered between the poles of an opposition. There is nothing natural, let alone justified, in proceeding from the richness of factuality to the utter refinement (or destitution) of a solid opposition, in which a contradiction is based less on its material content (which is stripped to the most basic elements) than on a division of a political nature. By 'political' we should not understand a division relating to external political positions (e.g. leftist or rightist), but *the most basic—and therefore inexplicable—division between positions, whose content is determined only later.*[17]

Therefore the figure of the face in philosophy—the main trick of 'the philosophy of the face' in general—should be construed as *a device which oversees the accumulation and discharge of negative energy through the process of simplification.* As such, it reduces the material content of reasoning to the basic grid of political relations. Under the guise of seeking the ultimate safeguards for ethical injunctions, reasoning is continually made impoverished. The concept operating as the

17 This notion will be later developed with reference to Carl Schmitt's thought.

figure of the face seems to administer a process in which oppositions become increasingly pronounced, material content is regrouped around opposites and speculative potential in local contradictions is channelled into the main opposition. Then the concept appears in person and claims the ground that it retroactively organised itself. Therefore, even the present, which appears as the central modality of the epiphany of the face, is staged. Through its separation from the preceding and subsequent discourse, the epiphany appears as present, but only through the predetermined structure of flow.

The second conclusion that might be drawn from Levinas' *Le Dire et le Dit* concerns the relation between the figure of the face and the fate of energy flow after its release. As we have already seen, the 'concept of the face'—if we may use this term—administered the process of simplification, that is, of accumulating potential and its final discharge. After this release, when the opposition becomes unconditional and essentially devoid of material content, the concept of the face governs the next process, which is even more crucial. Just as the body turns into bones, which theological devices later reanimate into flesh, so too does the concept of the face rebuild the material content of the opposition it had previously levelled. Now, however, the content is fully subject to the ultimate goal of this philosophical discourse.

The process of simplification, which once tore down the dialectical edifices to build a bare, radical contradiction of a political nature, now gives way to the process of restoration—which preserves all the negative energy condensed at the previous stage. Thus the philosophy of the face attempts to use the discharged potential to produce a concept which is itself charged with negative energy, though to serve new goals. In this sense, the philosophy of the face is a complex device which first levels the rich dialectical tissue of argumentation, strips it to its bare political relations, only to harness the condensed energy and use it to construct the key concept.

The radical and unconditional character of the ethical injunction in Levinas' thought has little to do with the alleged sublime of the encounter with the Other. Even less can it pass for a universal philosophical parable which leaves its addressees in mute admiration, as it is sometimes depicted. Although Levinasian philosophy has become surprisingly popular (in some regards it has been used to found a common *doxa* of late modern ethics), it is not as innocently ethical as it attempts to be. Its concept of the face—total responsibility—draws upon reservoirs of negative energy; contrary to Levinas' intentions, however, this manoeuvre cannot simply turn Heidegger's philosophy, which it opposes, into entirely ethical thought. Although the radical destitution of the subject is meant to safeguard the purity of total responsibility, the infrastructure which governs the production and spending of negative energy cannot be simply dissolved, let alone veiled by the overall positive conclusion of Levinas' ethical injunction.

This infrastructure constitutes the other side of (post)modern ethical reasoning which strives to construct an injunction through the accumulation and discharge of argumentative energy. Contrary to the clear and proud face of this philosophy, its key supplement cannot be masked. Perhaps the eventual futility of Levinas' injunction—which is so radical that it must arrive at a dead end of quasi-idealist inertia—practically de-faces the face structure of this type of philosophising.

Vectors of destitution

The face paradigm of philosophy brushes against a much more serious problem: the imagery of destitution, impoverishment and purification in modern and—if such a classification is accepted—postmodern thought. Inevitably, even the slightest attempt to outline the origin of this movement must exceed the capacity of this paper; it might be only briefly mentioned here.

The obvious point of reference for Levinas is Heidegger's thought; the primacy of ethics in Levinasian thinking hardly conceals its repositioning of priorities from Heidegger's philosophy. Just as the early Heidegger in *Being and Time* [Heidegger 1996][18] concentrates on the ontological problem of the fundamental difference between Being and entities—thereby positing an ontological view as a kind of new *prote philosophia*—so does Levinas attempt to give primacy to ethical thinking.[19] Despite this reversal and other obvious differences (such as Levinas' wealth of inspiration from Judaism, which Heidegger ignored, to put it mildly—to say nothing of his endlessly debated anti-Semitism), Levinas remains heavily dependent on Heidegger's legacy in at least one crucial aspect: *in the tendency to seek what is most fundamental and decisive in a given philosophical context.*

This tendency should not be misconstrued; it is far from leading both thinkers to assume simple underlying essences or *hypokeimena*. It is problematised, self-referential and richly dialectical. Nevertheless, it is fuelled by the intrinsically modern drive to perceive philosophy as afflicted by a serious crisis—a fundamental indebtedness to an unknown source—which might only be overcome through operating on the necrotic tissue of existing philosophical discourse. This drive is

18 Martin Heidegger, *Being and Time*, trans. Joan Stambaugh, SUNY Press, New York 1996.

19 See also the crucial critique of the relationship between Levinas and Heidegger in Derrida's *Violence and Methaphysics* [Derrida 2001, p. 167 f.]. Derrida rightly grasps Levinas' considerable indebtedness to Heidegger, hardly masked by the displacement of superseding ontology with ethics.

palpable at least as early as Kant's philosophy, which presented itself as a critique of previous metaphysics, whose insufficiency and futility had become manifest. Having taken this crisis for granted, Kant sought to posit a metaposition for his philosophy, from which previous metaphysical systems could be presented as blatantly usurpatory; by overcoming their approach, a new solid philosophical ground would be established [Kant 1910, pp. xvii–xxiv].

Since Kant, continental Western philosophy is, to a large degree, built on a permanent and radical crisis, which seems to call for a thorough re-formulation of the paradigms of philosophising. Often, as in the case of Marx, Nietzsche and Heidegger, philosophy has aimed to overcome the very structure of its discourse and to abandon the term 'philosophy' as inextricably linked to the false paradigm of the past.[20]

In this light, the tendency to seek what is most fundamental—which both Heidegger and Levinas share, albeit with different results—clearly appears in its link to the ongoing crisis of Western philosophy since Kant. *This tendency, in fact, perpetuates, amplifies and re-stages the crisis, bringing the problem and its solution to the threshold of indifferentiation.* It appears as the crisis and its overcoming, thereby shifting the very act of philosophising from its material content to a self-referential practice parasitising on the history of philosophy.

In this regard, Levinas' gesture, superseding ontology—which had allegedly ruled Western thinking since antiquity—with ethics has truly ambivalent consequences. On the one hand, this move seems to provide the ultimate safeguard for ethical reflection in philosophy. As such, it is manifestly opposed to Heidegger's infamous reluctance to account for the dimension of the ethical in his thinking. After World War II, when this particularity and Heidegger's brief entanglement in the Nazi regime (and, perhaps much more significantly, his postwar silence

---

20 The link between Levinas and the scheme of crisis/overcoming was obviously noticed by Derrida, who opened his *Violence and Metaphysics* with the following: 'That philosophy died yesterday, since Hegel or Marx, Nietzsche, or Heidegger—and philosophy should still wander toward the meaning of its death—or that it has always lived knowing itself to be dying (as is silently confessed in the shadow of the very discourse which *declared philosophia perennis*); that philosophy died *one day, within* history, or that it has always fed on its own agony, on the violent way it opens history by opposing itself to nonphilosophy, which is its past and its concern, its death and wellspring; that beyond the death, or dying nature, of philosophy, perhaps even because of it, thought still has a future, or even, as is said today, is still entirely to come because of what philosophy has held in store; or, more strangely still, that the future itself has a future—all these are unanswerable questions. By right of birth, and for one time at least, these are problems put to philosophy as problems philosophy cannot resolve'. [Derrida 2001, pp. 97–98].

on the subject, coupled with his disgraceful reticence towards the Shoah) began to seem more than mere coincidence, Levinas hastily attempted to salvage Heidegger's legacy from the long shadow of the philosopher's political entanglements. From this perspective, Levinasian ethics presented a good opportunity to review Heidegger's thought, in order to repair its basic shortcoming: the mute paralysis before the history of Being which renders all ethical components futile (as irreparably marred by metaphysical thinking).

Despite its incontestably rich argumentative ornamentation, Levinasian ethics is part and parcel of the very same problem which manifested itself so glaringly in Heidegger's aethical thought. The history of modern Western thought here returns to *the tendency to oversimplification, which, in its self-referentiality, attempts to repeat the impoverishment inherent in its intellectual heritage, and thus gain its own independent standing.* In this sense, Levinas' strategy exploits the same mechanisms that can be traced through philosophies from Kant to Heidegger. But here, much more visibly than in Heidegger's thought, the discourse of simplification is carefully staged in order to produce a negative energy to account for the unconditionality of the conclusive injunction.

In this light, Levinasian total responsibility is nothing less than intellectual blackmail—not in the sense of a cheap sentimental exhortation to self-sacrifice in the name of some vague Other, but on a much deeper level. The concept of total responsibility is based the mechanism of simplification, but without a deeper analysis of its nature and genesis. Thus the 're-ethicisation' of Heidegger's thought cannot be deemed a thorough reconsideration; it is a *manipulation with the veneer of this philosophy within the crisis/overcoming structure.* Consequently, in Levinas' thought the 'concept of the face'—total responsibility—is constructed as a *place* to which we are brought (the archetypal encounter with the face of the Other, which expropriates me[21]) and from which we are led in a carefully staged sequence. The search for what is the most basic and decisive has been shifted to the field of

---

21 A fragment from *Liberté et commandement* provides a good example of this staging: 'L'épiphanie de l'Absolument Autre est visage où l'Autre m'interpelle et ordonne de par sa nudité, de par son dénuement. Il m'interpelle de son humilité et de sa hauteur. Il voit, mais reste invisible et, par là même, s'absout de la relation où il entre, reste absolu. L'absolument Autre, c'est Autrui. Et la mise en question du Même par l'Autre est une sommation de répondre. Le Moi ne prend pas seulement conscience de cette nécessité de répondre, comme il s'agissait d'une obligation ou d'un devoir dont il peut décider; il est, *dans sa position même*, de part en part responsabilité. Et la structure de cette responsabilité montrera en quoi Autrui, dans le visage, interpelle de très bas et de très haut—en ouvrant la dimension même de l'élévation' [Levinas 1994, p. 66].

ethics. It is meant to provide a definitive ethical safeguard, but the *governing of flow* which we previously associated with 'the philosophy of the face' is preserved, and even expanded.

Moreover, Levinas' thought enters a characteristically modern double bind, to use Derrida's favourite category. The staging which governs and releases philosophical potential to produce a politically charged concept of total responsibility simultaneously affects the position of this philosophy in relation to its predecessors—predominantly to Heidegger's thought. Levinas plays on its lack of an ethical dimension in order to stage a reversal within the (dis)continuity of the history of philosophy. Naturally, the polemical arguments that he formulated must be appreciated. They cannot, however, overshadow the fact that the practical lack of ethical thought in Heidegger's philosophy is being juxtaposed with total responsibility. As such, *this lack is not analysed in its entanglement with the tendency towards simplification; it is immediately juxtaposed with the purported countermeasure—total responsibility—which, however, draws equally from this tendency*. Therefore, the split between Heidegger and Levinas is not a real difference, but a displacement which conceals their shared structural background.

The government of ethics

Obviously, one might come out and ask: even if Heidegger and Levinas share a fundamental, structural approach, why should we bother to untangle them if Levinas' philosophy offers a consistent, radical and seemingly unequivocal foundation for a new ethics? Perhaps the forces which account for Heidegger's blindness might be used profitably in a Levinasian approach? We might also dare to say that this may be the only way to channel them. If they are not to be incessantly confronted by ethics to middling effect, they might need to be harnessed to produce an ethical injunction.

The temptation to use the power of negativity for this seemingly positive goal is obviously hard to resist. Nevertheless, this sly alliance might prove even more dangerous than the blatantly aethical Heideggerian discourse. *It brings central mechanisms of modern philosophy into the field of ethics, thereby making it, in its entirety, the stakes in the game between crisis and overcoming*. This act foils and discloses the hidden possibility left by the glaring absence of the ethical in Heidegger's thought. Instead of turning this absence into a field of confrontation on modern philosophy's terms, it might have been seen as a starting point for reconsidering a stunning aspect of modernity: the characteristic inertia of being, whose core deflects the ethical. The work of negativity, which Levinas uses to defend ethics, should be analysed in itself, as the source of the destitution of the modern universe.

The structure of crisis/overcoming is based on a powerful scheme: first, the crisis of thought is acknowledged; the philosophy practised to date appears to be insufficient and, more importantly, rotted by negativity from within. Second, the very same negativity which troubled predecessors is now to be used consciously, in order to *acknowledge the irreparable damage that had been committed.* This sketches out a scheme for re-constituting thought after the crisis. This strategy has marked the key moments in modern Western philosophy, from Kant to Heidegger. Due to its dominance, the history of this philosophy turns into *a sequence of critical moments* whose (dis)continuity is the history of negativity at work. Nevertheless, since the project of overcoming boils down to a vain attempt to trace the future path of negativity, none of these moments can become a real pivot on which history might turn. Each preserves its own momentum; each is a mark of the same recurring strategy.

In this light, Levinasian ethics appears to be a missed opportunity. Not only is it embedded in the crisis/overcoming structure itself, it also ensnares ethics in the play of modern philosophy. Its structure might be called—to use a Foucauldian phrase [Foucault 2012]—*the government of ethics*, that is, a set of implicit rules governing the flow of negativity between particular concepts, so that it might be released through the concept of the face and used to found an ethical injunction.

We should not be deceived by the strength of Levinas' exhortation to total responsibility. It does not stem from the power of the ethical argument itself. With Nietzschean scepticism, it should be claimed that it does not differ structurally from every exhortation which was previously produced in the philosophical structure of crisis/overcoming. In truth, it is the very same paralysing exhortation that negativity addresses to previous thought, splitting the history of philosophy in two. The necessity of overcoming metaphysics that Heidegger formulates after his *Kehre* [Heidegger 2000] structurally matches the message of total responsibility in Levinasian ethics.

As in all previous moments of crisis/overcoming, a substantial historical background is conceived to present the *current decisive moment* as the historical breakthrough of philosophy, which is meant to leave its past behind and start a new life. Just as Nietzsche had Socrates, Plato and Paul for the *bêtes noires* of the history of philosophy, and Heidegger had Descartes, Leibniz and Hegel [Heidegger 1997], Levinas has his own: his *bêtes noires* are, on the one hand, 'Greek thinking' (whose boundaries were traced by Franz Rosenzweig [1988]) and, on the other, Heidegger. His portrayal is *very close to the truth*: indeed, it was in Heidegger that

Western philosophy demonstrated its radically aethical character. Nevertheless, this attribute does not determine the history of Western thought as such. It shows only how modern philosophy was ravaged by the power of negativity.

The concept of the ethical

The blindness Levinas offers is, in retrospect, a unique opportunity to rethink *the threshold at which a modern negativity-fuelled philosophy entered the domain of ethics*. This is not the moment when the government of ethics appeared as such (its features can be discerned back in Kant's and Nietzsche's thought), but when its problematic nature might have been grasped for the first time.

Levinas employed a careful argumentative device to channel the power of negativity into the strength of the ethical injunction. Unconditionality—so often appraised as a landmark of the Levinasian approach in ethics—is the work of negativity at its purest. The government of ethics, operating through the concept of the face, carefully arranges the natural-seeming persuasive power of the injunction. Therefore, *ethics becomes instrumentalised; its concepts and relations are disjointed and re-organised by the power of negativity*. Naturally, all ethics are biased by cultural and historical stereotypes; however, the government of ethics attempts to veil them in the process of discharging the accumulated flow. This is discernible in the above-mentioned example of Levinas' sexism: the usual concealment of inequality under the guise of naturalness is here not only considerably strengthened, it is also blocked by being mixed with the archetypal scene of the encounter with the Other.

The moment in which Levinasian ethics appeared on the scene of Western philosophy should be seen as pivotal. Under this assumption, Levinas' ethics must be meticulously re-assessed as a starting point for grasping the scope of the government of ethics. The ultimate horizon of this research should be guided by the hope that perhaps the total colonisation of ethics by the power of negativity provides a chance for resistance. Naturally, one should be careful not to reverse Levinas' gesture; seeking definitive salvation in extreme catastrophe is a recipe for repetition. The moment in which the government of ethics blatantly discloses itself should be seen as piercing through the repeating pattern of crisis/overcoming. Ethical thinking becomes dissolved and subjugated to the reign of deep, hollow structures. But in this state it still might be reinvigorated on another level. It might be used to reassess the work of negativity in modern philosophy, thereby paralysing the unconditional claims of the latter.

How might this be? The solid construction of ethics lies in ruins. Its claims are permeated with the underlying flows of energy which are organised by and around the concept of the face. *Yet this process has also a flip side: the negative energy of thought adheres to the remains of ethical thinking.* Ethical argumentation is dependent on its internal network of forces, but, for the first time in philosophy, these forces are closely linked to remnants of ethical concepts. Negativity establishes the unconditionality of ethics, but since it *can* establish it (just as it established other philosophical exhortations in the past), its power to produce unconditionality *might* also be questioned in the light of ethical thinking. Because even if the work of negativity is here to stay, *its usage*—namely the government of ethics—*might be subject to ethical reflection.* If the ethics which had hitherto been practised was based on *content*, not structural forces, its destruction calls into question the ethical status of these forces. This is the possibility that, contrary to Levinas' intentions, his ethics opens up.

The term 'ethics' does not seem strong enough to bear this new task. If the relation of forces which shape philosophical thinking (and ethics in particular) is to be analysed in this new light, perhaps a new term should be coined: *the ethical.* In order to explain it, we need to recall its counterpart: the concept of the political, formulated by Carl Schmitt.

Schmitt's notion of 'the political' names a division of a fundamental kind. It transcends all divisions based on content, such as that between good and evil (in the moral sphere) or beautiful and ugly (in the aesthetic sphere) [Schmitt 2007, p. 26]. It might accidentally coincide with them, but its function remains independent. In its groundlessness, the political fuels the ultimate passion-generating opposition; were it not for the political, there would be no true wars and no sacrifice of blood. Schmitt associates this fundamental division with the distinction between friend and foe [Schmitt 2007, p. 26]. This determination, however, does not offer an explanation for the nature of this division, it simply names its opposing poles. The nature of the political cannot be explained as such, since it is a fundamental division. Even less explicable is the nature of the 'enemy'—a category absolutely devoid of content, in some sense utterly accidental. One can be an enemy, but it is hardly determinable who will be named the enemy of a political community.

Leaving aside the convoluted history of this concept, it might be rightly claimed to be an expression of the modern tendency to simplify at its purest. No material justification can determine or change the nomination of the enemy once it is settled, and so, clearly, something other than a material rationale is at work here. This division is the foothold of negativity and, if sufficiently aggravated, its ultimate product. The tendency to reduce all material differences, to order and subjugate them to one, elementary, all-encompassing division, in which all of reality is encompassed by one either/or, *is structurally the same strategy which accounts for the government of ethics.* According to Schmitt, the political is anchored in the functions of the state; but the government of ethics, even if it has nothing to do with the state, structurally practices the same work of negativity which founded modern sovereignty. Therefore, it might be asserted that *Levinasian ethical reflection posits the encounter with the Other within the framework of the political.* This gesture brings the history of modern ethics to its culmination and its ultimate disintegration.

If there is to be any revival of ethics after this move, the field of ethical reflection must be extended to include the function of negativity and the problem of the political. Instead of ethics, perhaps we should think about *the ethical*: the type of philosophical enquiry which calls into question the work of negativity and its usage, particularly within the framework of the political. Old ethics is no longer possible without gross philosophical blindness: since ethics itself is beset by negativity, it cannot confront it. The ethical, however, provides a counterbalance to the political. Just as the political subsumes all differences into one, definitive category—thereby precluding all sensible reasoning as to its meaning—so the ethical must derail and foil this process. *The ethical questions each step of an argument as the potential work of negativity.* Naturally, negativity and the political cannot be eradicated; but their silent influence can be recognised and subjected to ethical reflection. In this way, the device epitomised by the 'concept of the face' might be dismantled.

Within the domain of the ethical we should analyse what ideas play the role of concepts of the face. No unconditionality should be taken for granted; on the contrary, each should be carefully vetted in order to disguise what material content has been welded to the very structure of the argument. The government of ethics should be dismantled along the lines of its political basis.

Modernity is so deeply linked with the functioning of negativity[22] that it is hardly imaginable that their paths might some day diverge. This does not mean, however, that we should succumb to its work, let alone sacrifice our ethics on the altar of the political. Levinasian ethics was a unique point in the history of Western philosophy, in which negativity engulfed ethical reflection under the guise of the most sublime injunction of responsibility. The figure of the face, so prevalent in Levinas' imagery, masks the indebtedness of ethics to the political; but, on the other hand, it simultaneously provides a trace of this unholy alliance.

---

22 The link between modernity and negativity was well grasped by Julia Kristeva in her pivotal book, *La revolution du langage poétique* [Kristeva 1974].

The disintegration of ethics by the political, even if irrevocable, might be a starting point for tales of *the ethical*. And perhaps it is the mirror of the ethical that could disturb the solitary and unconditional work of negativity.

Contrary to the Hölderlin's famed dictum, the greatest danger does not augur salvation. But in the stillness that it brings, we may discern, with unparalleled clarity, the ambiguity of our hopes.

Bibliography

1. Agamben, G., 2002. *Idea della prosa*, Quodlibet, Macerata.
2. Agamben, G., 1998. *Quel che resta di Auschwitz: L'archivio e il testimone*, Bollati Boringhieri, Torino (English edition: *Remnants of Auschwitz. The Witness and the Archive*, trans. D. Heller-Roazen, Zone Books, Cambridge, MA 2002).
3. Bauman, Z., 2001. *The Individualized Society*, Polity Press, Cambridge.
4. Derrida, J., 2001. 'Violence and Metaphysics', in: *Writing and Difference*, trans. A. Bass, Routledge, London and New York, pp. 97–192.
5. Foucault, M., 2012. *Du gouvernement des vivants. Cours au Collège de France. 1978–1979*, Seuil/Gallimard, Paris.
6. Heidegger, M., 1996. *Being and Time*, trans. J. Stambaugh, SUNY Press, New York.
7. Heidegger, M., 1993. 'Die Negativität. Eine Auseinandersetzung mit Hegel aus dem Ansatz in der Negativität', in: *Gesamtausgabe 68. Hegel*, Vittorio Klostermann, Frankfurt am Main.
8. Heidegger, M., 1997. *Gesamtausgabe 10. Der Satz vom Grund*, Vittorio Klostermann, Frankfurt am Main (English edition: *The Principle of Reason*, trans. R. Lilly, Indiana University Press, Bloomington, IN 1991).
9. Heidegger, M., 1983. *Gesamtausgabe 29–30. Die Grundbegriffe der Metaphysik. Welt—Endlichkeit—Einsamkeit*, Vittorio Klostermann, Frankfurt am Main.
10. Heidegger, M., 2000. 'Überwindung der Metaphysik', in: *Gesamtausgabe 7. Vorträge und Aufsätze*, Vittorio Klostermann, Frankfurt am Main (English edition: 'Overcoming Metaphysics', in: J. Stambaugh (trans. and ed.) *The End of Philosophy*, Harper & Row, New York, 1973).
11. Kant, I., 1910. *The Critique of Pure Reason*, trans. J.M.D. Meiklejohn, P.F. Collier and Son, New York.
12. Kristeva, J., 1974. *La revolution du langage poétique*, Seuil, Paris.
13. Levinas, E., 1971. 'Le Dit et le Dire', *Le nouveau commerce*, No. 18–19.

14. Levinas, E., 1974. *Autrement qu'être ou au-delà de l'essence*, Martinus Nijhoff, La Haye (English edition: *Otherwise than Being, or Beyond Essence*, trans. A. Lingis, Kluwer Academic Publishers, Dordrecht and Boston, MA 1978).
15. Levinas, E., 1987. *Hors sujet*, Fata Morgana, Montpellier.
16. Levinas, E., 1990. *Difficult Freedom*, trans. S. Hand, Johns Hopkins, Baltimore.
17. Levinas, E., 1991a. *Le temps et l'autre*, Quadrige/PUF, Paris.
18. Levinas, E., 1991b. *Totalité et infini: Essai sur l'extériorité*, Le Livre de Poche, Paris (English edition: *Totality and Infinity: An Essay on Exteriority*, trans. A. Lingis, Duquesne University Press, Pittsburgh 1969).
19. Levinas, E., 1994. *Liberté et commandement*, Fata Morgana, Montpellier.
20. Nietzsche, F., 1999. 'On Truth and Lying in a Non-Moral Sense', in: *The Birth of Tragedy and Other Writings*, eds. Raymond Geuss, Ronald Speirs, Cambridge University Press, Cambridge.
21. Rosenzweig, F., 1988. *Der Stern der Erlösung*, Suhrkamp, Berlin (English version: *The Star of Redemption*, trans. W. W. Hallo, Holt, Rinehart and Winston, New York 1971).
22. de Saint-Cheron, M., 2010. *Conversations with Emmanuel Levinas*, trans. Gary D. Mole, Duquesne University Press, Pittsburgh PA.
23. Schmitt, C., 2007. *The Concept of the Political: Expanded Edition*, trans. George Schwab, The University of Chicago Press, Chicago and London.
24. Žižek, S., 2014. *Absolute Recoil: Towards a New Foundation of Dialectical Materialism*, Verso, London and New York.
25. Žižek, S., 2006. 'Neighbors and Other Monsters: A Plea for Ethical Violence', in: Kenneth Reinhard, Eric L. Santner, Slavoj Žižek, *The Neighbor: Three Inquiries in Political Theology*, University of Chicago Press, Chicago, pp. 145–190.

Mathieu Corteel

Face 2.0. A philosophical approach to Facebook's semiotic facialization

This paper highlights the subjectification process in Facebook's semiotics. The aim is to explain the individuation, or production of identity, at work in this web machine. Through this study of Facebook as a machine that produces signs, an ontological structure emerges. The self's state of flux is produced by subjectification and intersubjectivity. The face is coded, decoded and overcoded by the flow of desire and gazes. The model of the face developed by Deleuze and Guattari in *A Thousand Plateaus* lets us give a precise explanation for this production of semiotic identities. This present paper is focused on a comparative study of the concept of faciality and the Facebook device. Furthermore, by questioning the concept of faciality in the age of new technologies, it leads us to explore posthumanism. From a philosophical perspective, this machine is a political device. With the face, posthumanist society defines a new model for recording a fluid identity, containing large amounts of data. Having a face is a societal injunction. We construct it through norms and values. Thereby everyone shares his or her personal data. Meanwhile, we are regulated by a latent power.

Keywords: Facebook, faciality, Deleuze and Guattari, individuation, posthumanism

Introduction

The face is a structure. It is a particular image of myself. It is also a common identity shared in the *Sittlichkeit*. A photograph of the face appears in every passport or official document. It is a key feature in a person's records. The face stands opposed to anonymity.

> I am no doubt not the only one who writes in order to have no face. Do not ask who I am and do not ask me to remain the same: leave it to our bureaucrats and our police to see that our papers are in order. At least spare us their morality when we write [Foucault 1969, p. 17]

This morality Foucault mentions controls the face. Through a logic of signification, authorities organise faces, as they do names. The order of appearance defines an ontological record of the self in the administration of life.

The deconstruction of this semiotic system permits us to highlight the structure of identity in a posthumanist society. From this perspective, I would like to deconstruct the semiotic system of the face, by examining the most well-known face delivery system: Facebook. In my point of view, this social network is not only a medium or media that facilitates communication between users, as we usually see it. It is more than that. It is a device, a machine that produces faces in a very specific way; in what Sloterdijk [2001] calls a 'posthumanist' way. Through facialization, Facebook converts the letter and the book of humanist selfhood into a massive production of signs. Using the rhizome of friendship, civil society is reconfigured by algorithms. And progressively, **signs conceal words**. Writing to lose one's face became marginal.

> Because of the formation of mass culture through the media – radio in the First World War and television after 1945, and even more through the contemporary web revolution – the coexistence of people in the present societies has been established on new foundations. These are, as it can uncontrovertibly be shown, clearly postliterary, postepistolary, and thus posthumanistic. Anyone who thinks the prefix `post' in this formulation is too dramatic can replace it with the adverb `marginal'. Thus our thesis: modern societies can produce their political and cultural synthesis only marginally through literary, letter-writing, humanistic media. Of course, that does not mean that literature has come to an end, but it has split itself off and become a *sui generis* subculture, and the days of its value as bearer of the national spirit have passed. The social synthesis is no longer – and is no longer seen to be – primarily a matter of books and letters. New means of political-cultural telecommunication have come into prominence, which have restricted the pattern of script-born friendship to a limited number of people. The period when modern humanism was the model for schooling and education has passed, because it is no longer possible to retain the illusion that political and economic structures could be organized on the amiable model of literary societies. [Sloterdijk 2001, p.3]

In this paper I will approach the problem of individuation – the passage from the substance to the subject – through a subculture of study abroad literature. The choice of Facebook as subject of study was made to deconstruct the faciality in posthumanist society. Through explaining the process inherent in the actualisation of faces, this paper highlights a general dynamic of identity in the new age of technology. From this perspective, the face as a system of signs stands in for the humanist construct of the self. The face as a construct of selfhood underlines the posthumanist dynamic of control and registration. The illusion of the self and its manifestation through entertainment dissimulates this aspect. The turn from a humanist to a posthumanist society underscores the administration of life as a human zoo. The entertainment of that system conceals a latent domination that we have to deconstruct. The study of the face must contribute to the deconstruction of the dominant culture. In this paper, by studying the construct of the face, I intend to bring out this new model of administration.

In my view, Facebook exemplifies a general prevailing dynamic of individuation in posthumanist society. I would assert that posthumanist individuation can be explained through its entanglement of the virtual and the actual. In other words, the thesis I will defend here is the relation between the virtual and the actual in the production of faces; the decoding and the overcoding of the face through actualisation.

Through individuation, a virtual device always defines a semiotic order. In Facebook, the semiotic transcends the face. This involves a description of the semiotic organisation. As such, I will leave aside the sociological and political facets to focus on one particular aspect: the semiotic production involved in the process of facialization. The difficulty of this approach is that the technology evolves every day. Tomorrow this article will be obsolete. In the empire of ephemera, theory itself is ephemera. Everything is in motion. Like our society, Facebook is not a fixed system. It is always changing. This is why I will try to approach Facebook dynamically, as a productive machine.

The process: virtual/actual

To begin with, I need to give precision to the movement that transcends this device. Conceptual precision on the concept of 'virtual' is necessary before we explain the semiotic system. The process of 'facialization' is comprehensible only if we explain the dimensions of individuation. Individuation is a condition which makes the face possible.

According to Gilles Deleuze [1968], the 'virtual' is not understood as the opposite of 'reality', but of the 'actual'. We often confuse this ontological relationship with the 'possibility/reality' opposition. The possibility/reality opposition science uses through the hypothetical-deductive method disrupts our representation of individuation. As Deleuze stresses, individuation is a product of multiplicity. It has a different formula.

In this formula, the actual and the virtual construct a scheme of differentiation. It is a product of simulacra. The relationship between the virtual and the actual is what we call 'actualisation', which is, according to Deleuze, a fundamental principle of differentiation. This principle of differentiation is inseparable from repetition, and comes from the multiplicity inherent in the virtual substance. Deleuze defines the virtual as 'a pure multiplicity in the idea' [Deleuze 1968, p. 273]. This definition allows us to discern between the virtual and the category of possibility, which Deleuze considers to be an 'identity in a concept'. This means that the virtual exists independently. It concerns the autonomy of the idea, and not the subordination of the concept. Possibility involves a concept, and depends on the thinking substance.

Therefore, we have two pairs: in pure reason, the 'virtual/actual' pair and, in understanding, the 'possibility/reality' pair. In this paper, I will only focus on the former pair, because of the flow of simulacra generated by the system I aim to explain. In this process, the virtual is related to the actual through repetition, through 'difference without concept' [Deleuze 1968, p.40]. This means that there is no limit to the production of signs. Actualisation as a process of individuation constitutes a powerful source of creativity. From this principle of differentiation we can deduce:

a) That the virtual exists through an infinite multiplicity. The relation to the particularity of the actual generates a creative production of simulacra.
b) That the virtual is the ontological status of the substance from which the individuation takes form through actualisation. As such, the process of actualisation generates difference from a multiplicity in perpetual motion. In brief, the virtual is the condition of possibility of actualisation, which facilitates perpetual updating. That is to say, the virtual survives the process. The substance is not spent by individuation.

This actualisation comes from the perpetual updating of the face on the network. This update is individuation in motion, with a rhizomatic structure. This means that updating reduces the multiplicity but does not expend it. It consists of a production of signs that generates an externalisation of the inner substance in an external structure: a landscape. According to Deleuze and Guattari, this construction of representation occurs in a deterritorialised space. Understood as a virtual machine in the rhizome of friendship, Facebook reports this condition of production. The face is stabilised through a rhizomatic structure in motion, and more importantly, this structure itself is mutating. Both the face and its platform are in motion.

Thus, we can conclude that individuation occurs through updates. 'The news feed' is the source of this movement. Consequently, the remainder of this study will investigate how the face is organised by the actualising and motion of this system.

The concept of the 'actualisation' has appeared as a metaphor in this social network, and on the Internet in general. My use of the concept is more extensive. It highlights not only the refreshing involved, but also the ontological structure defined by the relationship between subjectivity and the semiotic. As I understand it, Facebook has developed its own ontology of face production by actualising the multiplicity of the virtual. In the remainder of this study, I will explain this process with the concept of 'facialization' found in Deleuze and Guattari [1980], through a comparative analysis with Facebook.

The 'White-wall-black-hole' device

Initially, Facebook was created by Mark Zuckerberg to share such personal data as photographs, name, age etc. Self-representation has always been fundamental to this device. The self is an inexhaustible source of production. It is the raw material used by the system to produce its phantasm. The ego as substance is used to dream the world. In its protoform, Facebook was a trombinoscope of Harvard's student body. At that time, an avatar or a mask limited the manifestation of the self. The cybernaut appeared in the network through a totem, and not through a face. Everyone was represented by the Harvard totem. There was no overcode on individuation. The construction of the self was coded and limited by a totem. Later, this staticity was converted into a dynamic flow of representation. This marked the economic turning point of the company.

> This is an affair not of ideology but of economy and the organization of power (pouvoir). We are certainly not saying that the face, the power of the face (la puissance du visage), engenders and explains social power (pouvoir). Certain assemblages of power (pouvoir) require the production of a face, others do not. If we consider primitive societies, we see that there is very little that operates through the face: their semiotic is nonsignifying, nonsubjective, essentially collective, polyvocal, and corporeal, playing on very diverse forms and substances. This polyvocality operates through bodies, their volumes, their internal cavities, their variable exterior connections and coordinates (territorialities). [Deleuze and Guattari 1980, p.175]

The quality that has changed Facebook and spread it throughout the world was the virtual aspect of this device. When Facebook became deterritorialised it became an economic power. In other words, the movement permitted by actualisation has generated a dynamic trend of self-representation and money. Nowadays, the Facebook network does not use avatars and masks as totemic representations. It produces faces through a process of actualisation. From the primitive totemic territory to its capitalist schizophrenia, Facebook has evolved through the multiplicity in the idea. Facebook became a virtual machine producing facialization from a semiotic mechanism of information flow in perpetual motion.

The principle is easy to understand. It consists of a fluctuating web page set in motion by notifications, events, photos, post and comments. The rhizome of friendship lends significance to the flow of actualisation. I believe that the actualisation flow created through relations with our friends is Facebook's great invention. It generates an autonomous system of representation.

'The news feed' that actualises presence through a system of signs and subjectivity is Facebook's innovation and source of development. To illustrate this, in 2012, the network bought 'Instagram', a Smartphone application that shares photos directly onto Facebook. This updating of a photo in the 'news feed' changes the totemic aspect of the avatar and the mask in producing a face. What is the semiotic process that facilitates this production? The key explanation appeared in *A Thousand Plateaus* [1980], in which Deleuze and Guattari explained the semiotic production of the face. In the Chapter 7 of the book, entitled 'Faciality', they provided evidence for the systemic aspect of the face. In this chapter, the authors explain that faciality is produced by two sources of signification: the signifier and subjectivity (or the post-signifier). They call the entanglement of these two factors 'mur blanc-trou-noir' (white-wall-black-hole). They write as follows:

> Concrete faces cannot be assumed to come ready-made. They are engendered by an *abstract machine of faciality (visagéite),* which produces them at the same time as it gives the signifier its white wall and subjectivity its black hole. Thus the black hole/white wall system is, to begin with, not a face but the abstract machine that produces faces according to the changeable combinations of its cogwheels. Do not expect the abstract machine to resemble what it produces, or will produce. [Deleuze and Guattari 1980, p. 189]

The device described by Deleuze and Guattari is a pertinent way of conceiving Facebook. The abstract machine of faciality is exactly the same as Facebook's production of faces. We have the 'white-wall', where there is an actualisation of the signifier by the production of signs. In this space all signs are interdependent on other signs. Every sign is connected to the signifier through the network in a semiotic system. On this 'white-wall' every post is open to other posts. Meaning appears through the relationships between signs and users. The friendship system involving 'comments' or 'likes' entangles one with the other.

This forms a semiotic system in a virtual space. Actualisation is the principle of this program. It structures these connections of signs. Thereby, we can modify a post or a comment, and also create other posts in the flow of self-representation. Hence, we have what is called a 'black-hole'. This is the multidimensionality generated by our subjectivity and the intersubjectivity stimulated by the flow of data. The black-hole makes connections between 'news feeds'. It is the line of flight that connects everyone. Subjectification feeds the system. Individuation appears on a subjective background and makes the system itself flexible. This is why Facebook is not only a medium focused on a present reality. In my point of view, it is a virtual machine producing simulacra through simulating self-representation.

> The simulacrum is never that which conceals the truth – it is the truth which conceals that there is none. The simulacrum is true. [Baudrillard 1981, p. 9]

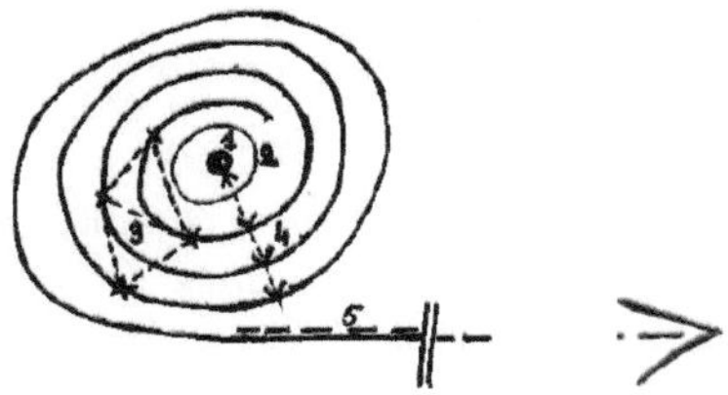

This structure by Deleuze represents the entanglement of the two different types of signs. The signifiers (white-wall) are represented by the spiral (points 1 to 4) and 'post-signifying' subjectivity (black-hole) by the line of flight (points 5, 6….) The white-wall generates an accumulation of strata and the black-hole expands the connection between multiple spirals.

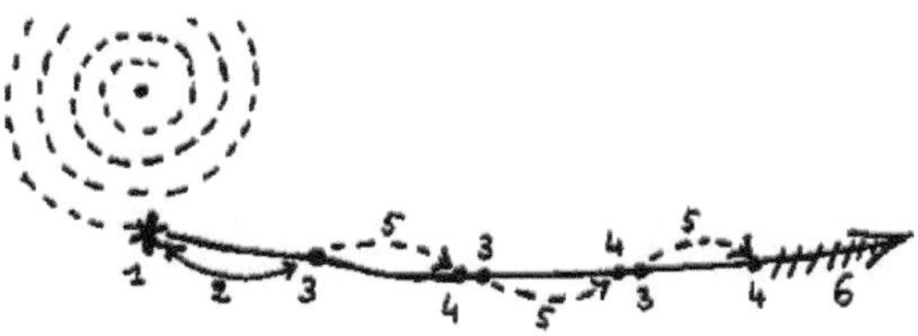

(1) The point of subjectification, replacing the center of signfiance. (2) The two faces turned away from each other. (3) The subject of enunciation resulting from the point of subjectification and the turning away. (4) The subject of the statement, into which the subject of enunciation recoils. (5) The succession of finite linear proceedings accompanied by a new form of priest and a new bureaucracy. (6) The line of flight, which is freed but still segmented, remaining negative and blocked.

[Deleuze and Guattari 1980, pp. 135–37]

The decoding and the overcoding of the body (the white-wall semiotic device)

According to Deleuze and Guattari, the face is defined by a system of surface and hole. It is a map, not a territory. The face is the informational facet of the human being. It is inwardness externalised through an informational component. This is why the Face is not the body, and why we do not confuse it with the human head. It is not a proprioceptive system, it is an informational system.

> The head, even the human head, is not necessarily a face. The face is produced only when the head ceases to be a part of the body, when it ceases to be coded by the body, when it ceases to have a multidimensional, polyvocal corporeal code—when the body, head included, has been decoded and has to be *overcoded* by something we shall call the Face. [Deleuze and Guattari 1980, p. 191]

The face is inscribed in a process of disembodiment by the white-wall. It is an entity that overcodes the body by producing signs from a signifier. The code of the body – the territoriality of the organic system – has to be deterritorialised to create a face. The body has to disappear to let the face appear on the device. As such, the whole body has to be decoded and overcoded on a face.

> Hand, breast, stomach, penis and vagina, thigh, leg and foot, all come to be facialized. Fetishism, erotomania, etc., are inseparable from these processes of facialization. It is not at all a question of taking a part of the body and making it *resemble* a face, or making a dream-face dance in a cloud. No anthropomorphism here. Facialization operates not by resemblance but by an order of reasons. It is a much more unconscious and machinic operation that draws the entire body across the holey surface, and in which the role of the face is not as a model or image, but as an overcoding of all of the decoded parts. [Deleuze and Guattari 1980, p. 191]

The decoding and overcoding of the proprioceptive body produce a part of face on the white-wall. This production projects the body's inwardness on the exterior as an informational manifestation of the signifier. It is a deterritorialisation from the body to a reterritorilised landscape. The overcoding fixes the signifier in a representative system of signs. A landscape is connected to stabilise the face in a striated territory that gathers strata of representation. In other words, the face is stabilised by a semiotic space that accumulates data. In Facebook's case, this is cyberspace, a mental geography.

> Cyberspace: A common mental geography, built, in turn, by consensus and revolution, canon and experiment; a territory swarming with data and lies, with mind stuff and memories of nature, with a million voices and two million eyes in a silent, invisible concert of enquiry, dealmaking, dream sharing, and simple beholding.
>
> Cyberspace: Its corridors form wherever electricity runs with intelligence. Its chambers bloom wherever data gathers and is stored. Its depths increase with every image or word or number, with every addition, every contribution, of fact or thought. Its horizons recede in every direction; it breathes larger, it complexifies, it embraces and involves. Billowing, glittering, humming, coursing, a Borgesian library, a city; intimate, immense, firm, liquid, recognizable and unrecognizable at once.
>
> Cyberspace: Through its myriad, unblinking video eyes, distant places and faces, real or unreal, actual or long gone, can be summoned to presence. From vast databases that constitute the culture's deposited wealth, every document is available, every recording is playable, and every picture is viewable. Around every participant, this: a laboratory, an instrumented bridge; taking no space, a home presiding over a world ... and a dog under the table. [Benedikt 1991, p. 2]

This mental geography called cyberspace is without a doubt the place where the overcode takes shape and informs the semiotic 'white-wall'. Thus, the decoding of the body and the overcoding of the face establish the signifier's structure. The mental geography lets the body vanish to let the face appear. The simulacra are aggregated on the face like the shifting of tectonic plates. The accumulation of strata reinforces the illusion of self-representation. The map progressively overtakes the territory. It generates expansion through the signifier's mode of expression: updates. Everyone produces these when they express themselves on the network. When we post something onto the white-wall, we expand the illusion. The production of data through a face is the source of this kind of space.

> … In that Empire, the Art of Cartography attained such Perfection that the map of a single Province occupied the entirety of a City, and the map of the Empire, the entirety of a Province. In time, those Unconscionable Maps no longer satisfied, and the Cartographers Guilds struck a Map of the Empire whose size was that of the Empire, and which coincided point for point with it. The following Generations, who were not so fond of the Study of Cartography as their Forebears had been, saw that that vast Map was Useless, and not without some Pitilessness was it, that they delivered it up to the Inclemencies of Sun and Winters. In the Deserts of the West, still today, there are Tattered Ruins of that Map, inhabited by Animals and Beggars; in all the Land there is no other Relic of the Disciplines of Geography. [Borges 1960, p. 40]

The post-signifier, the subjectivity of the black-hole

With the post-signifier, the network is perpetually in motion, returning to the virtual aspect of the device on lines of flight. Why does this return from actual to virtual take place through the individuation process? It is because the multiplicity generated by the community of friends (the system of friendship) destabilises the strata expansion of the white-wall. Through repeating the 'black-hole' as a line of flight, every self is connected to the network. A connection of subjectivity, a link between substances, is created from a friendship, which expands the representation. This particular link, forged from shared passions, develops in the flow of its manifestation. This is what Deleuze and Guattari call a 'point of subjectification'.

The black-hole is a point of subjectification. It, is, for example, a dress or shoes for a fetishist. It is a line of flight that abstracts objects of desire through passions. On this virtual device, the rhizome of friendship always elicits our subjectivity. The Facebook news feed stimulates us by lines of flight that come to exist through actualisation. We have to be active to exist. Thus, the 'news feed' is the motor that generates passions through a virtualisation of the self and an actualisation of the consciousness. This semiotic production is what Deleuze and Guattari have called a 'post-signifying semiotic', in which subjectivity is manifested through the passions in lines of flight.

> The postsignifying semiotic in which overcoding is assured by the redundancy of the consciousness; a subjectification occurs on a passional line that makes the organization of power immanent and raises the deterritorialization to the absolute [Deleuze and Guattari 1980, p. 135]

The individuation provided by the post-signifying semiotic establishes the deterritorialisation of the white-wall by the repetition process of consciousness.

We definitively forget the body and the whole becoming-animal in order to exist on the network. The repetition of the consciousness in which subjectification occurs deletes the body and its affects by constructing passions. I believe that this happens in the network through a 'normative ideology'. Much like fetishism, desire appears through a point of subjectivity (shoes, a dress etc.). This point of subjectivity is a normative vector of desire. Hence, the post-signifying semiotic generates a collective layout that distributes the norm of regulation. In Facebook, it appears through the 'like button' function.

> The abstract machine of faciality assumes a role of selective response, or choice: given a concrete face, the machine judges whether it passes or not, whether it goes or not, on the basis of the elementary facial units [...] At every moment, the machine rejects faces that do not conform, or seem suspicious. [Deleuze and Guattari 1980, p. 177]

In fact, this show of friendship constructs a collective and particular individuation in a normative way. There is a trivial and biunivocal relationship established by the repetition of the 'like button'. When we use the 'like button' we repeat the subjectification of desire. It is a negative deterritorialisation in which desire is reduced to a point of subjectification. Moreover, the abstract machine controls the facial units. It administrates the organisation of subjectification.

This means that there is a deterritorialisation of substance in a line of flight that produces individuation in the rhizome of friendship. This individuation is tied to the decoding and the overcoding of the face on the device. The white-wall-black-hole generates an infinite simulacrum system, producing the veil of Maya. The rhizome of friendship generates negative lines of flight through the virtual 'black-hole' device. The black-hole deterritorialises the self through repetition. 'The white-wall is always expanding and the black-hole functions repeatedly' [Deleuze and Guattari 1980, p. 177]. We can say that this device clearly actualises faciality through the expansion of the white-wall and the repetition of the black-hole. Hence, this system generates an accumulation of facial strata. It is a spiral with lines of flight deterritorialised to form a machine with multiple eyes. This machine produces stigmergic traceability through collective individuation.

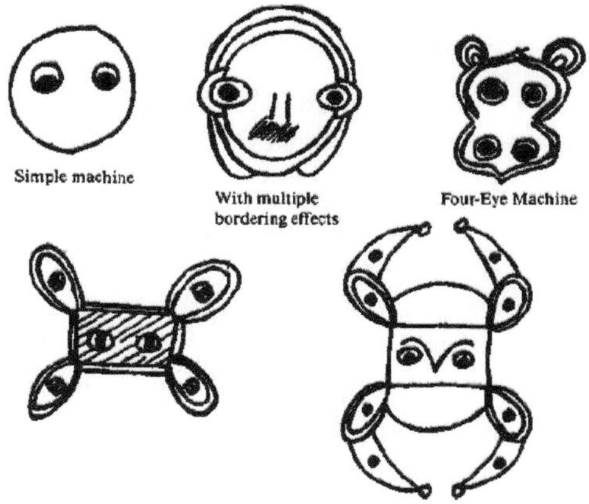

Proliferation of Eyes By Multiplication of Border

Terrestrial Signifying Despotic Face

This scheme represents the expansion of the black-hole as a multiplication of eyes. Through lines of flight the machine connects a multiplicity of eyes and modifies its own structure.
[Deleuze and Guattari 1980, p. 183]

The control of the face

Finally, there is a system of control that operates through faciality. The abstract machine of faciality develops a power system that imposes 'signifiance' and subjectification while it rejects the body and its organic expression. According to Deleuze and Guattari the face is political. The face rejects the body's affects through imposing signifiance and subjectification.

> *Very specific assemblages of power impose signifiance and subjectification* as their determinate form of expression, in reciprocal presupposition with new contents [...] A concerted effort is made to do away with the body and corporeal coordinates through which the multidimensional or polyvocal semiotics operated. Bodies are disciplined, corporeality dismantled, becoming-animals hounded out, deterritorialization pushed to a new threshold—a jump is made from the organic strata to the strata of signifiance and subjectification. A single substance of expression is produced. The white wall/black hole system is constructed, or rather the abstract machine is triggered that must allow and ensure the almightiness of the signifier as well as the autonomy of the subject [...] This machine is called the faciality machine because it is the social production of face, because it performs the facialization of the entire body and all its surroundings and objects, and the landscapification of all worlds and milieus. The deterritorialization of the body implies a reterritorialization on the face; the decoding of the body implies an overcoding

> by the face; the collapse of corporeal coordinates or milieus implies the constitution of a landscape. [Deleuze and Guattari 1980, pp. 180–81]

As indicated above, the abstract machine of faciality is a power system in which the strata of the body disappear to let stigmergic traces of faciality appear. I believe that this opposition to the becoming-animal is a way of ensuring traceability in society. I mean to say that this production of signs through the expression of the signifier and subjectivity define a regime of surveillance. This is a cognitive surveillance in which the body is left behind, and the face is controlled by monitoring devices.

The trivial aspect of 'selective responses' helps us to understand a particular kind of supervision deriving from an organisation of the system made by the 'EdgeRank' algorithm. This process consists in a certain disposition to the news feed. With users' participation in the news feed, the algorithm determines what articles should be displayed. It produces a *feedback* loop that generates interaction.

From the friendship link, a classification made by the algorithm is permitted to organise selective responses. The Facebook configuration administrates the network through the cooperation of individuals. This is a simulation of the society and the State. Through simulacra and strata (semiotic) the system demonstrates that there is no truth in the creation of a posthumanist society. The stratification of identity produces its own system of representation and power. A brave new world posing as worldwide fiction appears to dissimulate the construct of representation under the power of a semiotic system – insomuch as policy controls individuals by this system. The system produces its own objects of surveillance. Facebook is a pure apparatus of capture.

> What forms the apparatus of capture are two operations always found in the convergent modes: direct comparison and monopolistic appropriation [...] This is a white wall/black hole system of the kind that, as we have seen, constitutes the face of the despot. A point of resonance circulates in a space of comparison and constitutes that space as it circulates. That is what distinguishes the State apparatus from primitive mechanisms, with their noncoexistent territories and nonresonating centers. What begins with the State or the apparatus of capture is a general semiology that overcodes the primitive semiotic systems. Instead of traits of expression that follow a machinic phylum and wed it in a distribution of singularities, the State constitutes a form of expression that subjugates the phylum: the phylum or matter is no longer anything more than an equalized, homogenized, compared content, while expression becomes a form of resonance or appropriation. Apparatus of capture—the semiological operation par excellence... [Deleuze and Guattari 1980, pp. 444–445]

Conclusion

The individuation created by a virtual system like Facebook facilitates the constant production of signs through the semiotic of the signifier and the lines of flight of subjectivity. The constant actualisation generates a regime of power that is inescapable, because it is coextensive with individuation. In a liberal society everyone seeks autonomy and wants to express it through technology. That is why this coercion is latent and pervasive. It is a soft surveillance that gives the impression of freedom. Nowadays, with the development of predictive analytics from Big Data, the stratum of traces allows us to predict and anticipate our social behaviour and actions.

But what exactly does it anticipate? As I have mentioned, the system works based on feedback loops. In other words, the system repeats a model. It is a tautology. The model comes at the beginning and at the end. For everything produced in Facebook modifies the model itself. The dynamic structure evolves through the accumulation of strata and points of subjectificaton. This new capability offers a better way to monitor production to the abstract machine of faciality. In fact, this anticipation develops the system by constantly expecting individuation. It is a fabric of individuality in which control operates through expression. Everything is constructed. When we use this kind of technology to administrate life, we produce our own subjection. The simulacrum dissimulating the absence of truth becomes real. The social order developed by Facebook comes from the normative capability of the network.

EdgeRank organises the system through the users. Without users there is no truth. The device must have the raw material of self-representation. The psychogenesis produced by Facebook reveals the posthumanist selfhood as a productive machine.

> It is at work everywhere, functioning smoothly at times, at other times in fits and starts. It breathes, it heats, it eats. It shits and fucks. What a mistake to have ever said the id. Everywhere it is machines—real ones, not figurative ones: machines driving other machines, machines being driven by other machines, with all the necessary couplings and connections. An organ-machine is plugged into an energy-source-machine: the one produces a flow that the other interrupts. The breast is a machine that produces milk, and the mouth i machine coupled to it. The mouth of the anorexic wavers between several functions: its possessor is uncertain as to whether it is an eating-machine, an anal machine, a talking-machine, or a breathing machine (asthma attacks). Hence we are all handymen: each with his little machines. For every organ-machine, an energy-machine: all theime, flows and interruptions. Judge Schreber has sunbeams in his ass. A solar anus. And rest assured that it works: Judge Schreber feels something, produces something, and is capable

> of explaining the process theoretically. Something is produced: the effects of a machine, not mere metaphors. [Deleuze and Guattari 1972, pp. 8–9]

A connection occurs in the system. Libidinal connection, erotomania and fetishism participate in the process as points of subjectification. The machine evolves, connecting everyone through phantasms. 1.49 billion people use Facebook. The system is used in 137 out of the world's 193 countries. The map will surpass the territory until geography becomes a thing of the past.

Bibliography

1. Baudrillard, J., 1981. *Simulacres et simulation*, 1th Ed. Paris, Galilée.
2. Borges, J. L., 1960. *El Hacedor*, 11th Ed. Madrid, Alianza Editorial.
3. Benedikt, M. (ed.), 1991. *Cyberspace first steps*, Cambridge, Massachussets, The MIT Press.
4. Deleuze, G., 1968. *Différence et répétition,* 1th Ed. Paris, PUF, Epiméthée.
5. Deleuze, G., Guattari F. (1972), *Anti-Oedipus, Capitalism and Skyzophernia 1*, 10th Ed. Minneapolis, Minnesota, University of Minnesota Press.
6. Deleuze, G., Guattari, F., 1980. *A Thousand Plateaus, Capitalism and Skyzophrenia 2*, 11th Ed. Minneapolis, Minnesota, University of Minnesota Press.
7. Foucault, M., 1969. *The Archeology of Knowledge*, 1th Ed. New York, Pantheon Book.
8. Sloterdijk, P., 2001. 'Rules for the human Zoo: A response to the Letter on Humanism', in: *Environment and Planning D: Society and Space* 2009, Volume 27, pp. 12–28.

Anna Rowińska

The physiognomic lie of hysteria

In the Salpêtrière hospital for the mentally ill, Doctor Jean-Martin Charcot (1825–1893) used physiognomy as a method for constructing diagnostic tables of hysteria, an invented illness. Photographers took portraits of suffering women; their expressions and gestures served as a basis for a typology. These faces of hysteria were backed by the nineteenth century belief in the objectivity of photography. The drama and the visuality of hysteria were perfect for creating physiognomic diagnostic tables. This article offers a critical analysis of these photographs and the paradigm of the madwoman.

Keywords: photography, physiognomy, hysteria, typology

Hysteria has been a phenomenon in the history of art and culture since Antiquity, becoming the most characteristic illness of the *fin de siècle*. The nineteenth-century French neurologist Jean-Martin Charcot (1825–1893) carefully described and documented the psychological abnormalities he believed to be a neurological sickness he called *hysteria*. He manipulated his patient's physiognomies in order to support his medical theory. Photography became Charcot's main tool; it became the medium which revived the lie of hysteria. His thesis was debunked by his students, but it partly survived in psychiatry, psychology and neurology. The images created during the process are now a chapter in the history of art [Didi-Hubermann 2003], but they also bear witness to and produce lies and the manipulation of physiognomy for medical and artistic purposes.

The depiction of madness has a long tradition in the arts and culture. The paradigm of the madwoman has undergone various visual changes. However, there have been some recurring features: woman has been seen as a beast, black, hairy, wild and brave, excluded from society, situated on the border between the visible and invisible world. She was a confidante of the devil [Gilman 2014]. Women were portrayed differently from men: they were sinful and close to nature. The tradition of depicting madwomen is an important influence on Charcot's work. Clearly, he drew his ideas on mental illness from archetypes which were still very strong in the nineteenth century.

It is believed that the modern approach to mental illness – creating typologies, diagnostics and seeing the importance of therapy – began in the early nineteenth century, but it was the segregation initiated in the mid seventeenth century in France that made this possible. An edict of April 1656 founded the General

Hospital of Paris. Its main role was to take the poor or otherwise marginalised out of the city and into institutions in the suburbs of Paris [Foucault 1998]. Societal outcasts both healthy and sick were provided with care, but also received moral instruction. The eighteenth-century asylums were transformed into purely medical institutions [Shorter 2005]. These institutions facilitated continuous observation and monitoring. Methods of treatment often relied on brutal practices, and patients were kept in terrible conditions. Despite this, Michel Foucault described the phenomenon taking place between the seventeenth and the nineteenth century as a process 'from exclusion to liberation' [Foucault 1998, p. 23].

The era of the Enlightenment brought changes in how the mentally ill were both imagined and perceived, though there was still no single universal concept of madness. Sciences like psychiatry, psychology and neurology were born. The nineteenth century was a time when medicine developed rapidly, alongside industrialisation and urbanisation in Europe. These changes significantly altered the treatment and perception of madness. The most important, pioneering centres for the mentally ill were founded in France and England. Asylums were slowly transformed into hospitals that provided medical treatment.

Hysteria has a medical and mythological tradition. The oldest attempts to describe the illness can be found in the writings of Hippocrates and Galen. Its original form assumed that there was a link between the work of the female reproductive organs and attacks of insanity. Hysteria was considered a disease of the uterus. For centuries it was believed that, due to malfunction of this body part, women succumbed to madness [Jakubik 1979]. This opinion was shared by scientists of other eras, such as: Edward Jordan, Robert Burton, Ambroise Pare and Wilhelm Griesinger [Scull 2009]. In the Middle Ages hysteria was seen as synonymous with possession. Abnormal behaviour was interpreted as a problem of the soul, qualified as a crime or sin; hysteria became linked with immorality [Foucault 1998]. This view survived to the end of the eighteenth century.

Mysterious diseases fired the imaginations of societies on the verge of modernity – it is common to speak of the nineteenth century as the era of uteral diseases [Corbin et al. 2013]. However, after the Renaissance, some researchers expressed the view that hysteria was a disease of the nervous system, unrelated to genital dysfunction. The first to express this idea was Charles Lepois, followed by Thomas Willis; these two trends, the genital and cerebral origin of hysteria, coexisted until Charcot [Jakubik 1979]. Recognition of the brain being the locus of disease resulted in the statement that men, too, could suffer from hysteria. Charcot was proud of his comprehensive research on this subject [Gilman et al. 1993].

Despite detailed analyses, hysteria has always been associated with epilepsy and spectacular convulsions. Paradoxically, this brain illness or soul dysfunction was considered to be a visual phenomenon. Charcot decided that sickness was to be described mainly through observation.

Hysterics became the paradigmatic 'Other' of the nineteenth century. Society was both terrified and fascinated by hysteria, associating it with the release of dark forces. 'This disease, which influenced the image of a woman from the late eighteenth century, until the triumph of Freud's theory, is perceived primarily as a rhetoric of the body tormented by instincts and passions' [Corbin et al. 2013]. Charcot stormed into this world with his scientific methods and camera, which had an ambivalent function: it demythologised the demonic origins of the disease, but also confirmed its existence in a form which evoked fear and curiosity.

'Otherness' is the most important element of photography. Susan Sontag mentioned this category as fundamental in immortalising people [Sontag 2009]. From the very beginning, the margins of society were of major interest to photographers. As André Rouillé wrote: '*the Other* was included in the dialog of the photographic process.' [Rouillé 1989]. Madwomen were always an interesting topic, and with the invention of photography, their secrets could be captured, analysed, and the fascination spread.

Salpêtrière hospital outside Paris was the scene of revolutionary events in neurology and psychiatry. It came to exist in 1670, primarily as a shelter, repeatedly changed its function, and by the early nineteenth century it was famous as a hospital for the insane. It was a place devoted to women; not only the mentally or physically impaired were brought here, but also the poor, or otherwise excluded from society [Shorter 2005]. Because of the accumulation of people, their state and the conditions in which they were held, critics called the hospital the 'city of crazy women', 'women's Mecca of death', and 'hell' [Didi-Huberman 2003]. Salpêtrière went down in history as the birthplace of modern psychiatry; many famous doctors and theorists were associated with this institution, such as Philippe Pinel, Sigmund Freud or Gilles de Tourette.

Jean-Martin Charcot was born in Paris in 1825. He was a painter, and later used his talents in his medical research. He received his higher education from the School of Medicine in Paris, after which he started his practice in Salpêtrière hospital. In 1953 he became the institution's director [Bogousslavsky & Boller 2013]. He was also educated as a pathologist. Hysteria was his secondary interest (from 1870 onward), and he became famous for other respected discoveries: he was the first to diagnose a living patient with multiple sclerosis, isolated amyotrophic lateral sclerosis and co-founded a group of scientists who isolated a hereditary neuropathy called Charcot-Marie-Tooth diseases. He is considered to be the father of neurology, primarily through creating a cerebrovascular description and discriminating between tremors in Parkinson's disease and those occurring as a result of sclerosis. He began clinical diagnostics, rehabilitation, and created the diagnostics for cholangitis. Charcot distinguished hysteria from epilepsy, which made it possible to clarify the symptoms [Goetz & Bonduelle 1995]. He illustrated his writings and publications with his own sketches.

In the second half of the nineteenth century, there were around 4,500 inhabitants in Salpêtrière [Didi-Huberman 2003]. Charcot was fascinated by this accumulation of patients, as it improved the potential for his comparative method of diagnostics [Bogousslavsky & Boller 2013]. A large number of patients meant diversity and a large cross-section of cases, facilitating the creation of statistics on abnormalities in a wide population. Comparison gradually became the most important part of Charcot's work. This might have resulted from his interest in physiognomy, as it was deeply rooted in comparing cases and drawing conclusions based on differences and similarities.

Charcot wrote many publications. His most important titles on hysteria are: *Iconographie photographique de la Salpêtrière* (published every year from 1876 to 1880, written with photographer Paul Regnard and psychiatrist Désiré-Magloire Bourneville, whose last volume contained photographs by Paul Richer and typology boards), *Nouvelle iconographie photographique de la Salpêtrière* (1888, written with Gilles de Tourette, Paul Richer and Albert Londe) and *Les difformes et les malades dans l'Art* (with Richer). *Leçons sur les maladies du système nerveux faites a la Salpêtrière par J.-M. Charcot* consists of four volumes published from 1873 to 1877, and features neurologists' lectures collected by Bourneville. During Charcot's tenure at the hospital, other books were published, written by Londe, Richer, Bourneville or de Tourette. These writings also provide extensive information on the illnesses studied at Salpêtrière, on Charcot and his methods, and on photography – especially the writings of Londe.

In examining the patient (no photographs were taken during this stage), Charcot placed a great deal of emphasis on empiricism and visuality. This is how Georges Guillain describes his first study of a patient, occurring after his arrival at the hospital: 'He sits down near a bare table, and immediately has the patient to be studied brought in. The patient is then completely stripped. The intern reads the "observation", while the Master listens attentively. Then there is a long silence during which he gazes; he gazes at the patient and drums his fingers on the table. The assistants are standing, crowded together, anxiously awaiting a word that will shed some light. Charcot remains silent. Then he instructs the patient to move in a certain way, makes her speak, asks for her reflexes to be measured, for her sensitivity to be examined. And again he falls silent, Charcot's mysterious silence. Finally he brings in a second patient, examines her like the first, calls for a third, and, still without a word, compares them' [Didi-Huberman 2003, p. 22]. The neurologist categorised and analysed the phenomena of clinical types, described examples of illnesses, made observations and conducted research to find their anatomical basis. He emphasised the visual form of mental dysfunction. For Charcot, observation was less an intimate narrative of pathology than it was a well-made description of the state of the body.

It is important to note the use of the word *Iconography* in the titles. This illustrates that Charcot believed his new typologies to be both scientific and artistic. In the first half of the nineteenth century, with the growing interest in the religious art of the Middle Ages, the first iconographic books had begun to appear; they functioned as lessons on topics in art [Didi-Huberman 2003]. With the use of the word *Iconographie* Charcot accentuated the link between art and his theories. Didi-Huberman noticed a strong 'iconographic impulse' in Charcot, the imperative to structure in accordance with scientific and empirical criteria, while drawing reference to categories in art [Didi-Huberman 2003].

Didi Hubermann wrote an extensive book on Charcot's work, titled *Invention de l'Hystérie: Charcot et l'Iconographie Photographique de la Salpêtrière*. Although Charcot truly constructed, or 'invented' an image of hysteria, several scientists and philosophers had already created similar theories on the subject. Charcot's theory of hysteria was initially popular and respected, but was cast into doubt by the psychiatric society, students and even Charcot himself. Immediately following the neurologist's death in 1893 it was concluded that the whole phenomenon was a deception. Hysteria was the invention of one man who tried to build a scientific theory based on physiognomy and photography.

Charcot postulated that hysteria was a disease of the nervous system unrelated to the genitals, manifesting itself in attacks or permanent impairment. Charcot used existing traditions to distinguish between separate hysterical diseases, trying to embed them in neurology (not psychiatry or psychology, as it is often erroneously believed). His work was later adapted and continued by his student, Sigmund Freud. Autopsies were conducted on hysterics at Salpêtrière, and although they showed no physical disorders, this did not stop Charcot from postulating that the illness has a neurological substratum. Charcot distinguished several causes of the abnormalities, and although the trauma thesis prevailed, he also took into consideration poor nutrition and other factors [Charcot, Bourneville & Babinski 1892].

He believed that attacks and paresis physically 'replayed' an event that caused the illness. It might be said that hysteria was a memory imprinted in the body.

Charcot listed and illustrated several hysterical stigmata, as well as the most important elements of a hysterical attack, which he called *grande hysterie*. This structured and dynamic state of mind and body is what went down in history as Salpêtrière hysteria. The stages were (in order of appearance): epilepsy, when women mimicked an epileptic attack (it was said that they copied what they observed while living with epileptics in the same ward); *clownism*, which was characterised by illogical, chaotic gestures; *attitudes passionelles*, which was the phase of ecstasy; and finally, delirium. Before the attack, the patient may have sensed a peculiar aura [*aura hysterica*], characterised by palpitations, coughing, yawning, and obstruction of the throat [Charcot, Bourneville & Babinski 1892]. An attack consisting of these stages forms our popular notion of hysteria, but Charcot also singled out hystero-epilepsy and other types of hysterical contractures and deformities of the body. Polish psychiatrist Andrzej Jakubik pointed out that Charcot's hysteria mimicked a variety of organic diseases. He emphasised the role of emotions and suggestion in the formation of hysterical symptoms [Jakubik 1979].

In Charcot's publications we find photographs which illustrate each stage of a *grande hysterie* attack, other manifestations of hysteria and several other types of illnesses. Physiognomy plays a great role in these illustrations, as the face reflects madness, pain, suffering, ignorance and despair. Comparative plates and drawings of details accompany the photographs. Like Lavater, Charcot drew conclusions based on his own observations. Interestingly, not only did he compare drawings or photographs, but it is said that he often compared two or more living patients. His diagnostics were only based on observation, the role of conversation was secondary, and paradoxically, served more as an illustration to the visual element.

In the nineteenth century physiognomy became a basic methodological tool for presenting and diagnosing mental illness. This was the golden age of physiognomical and phrenological theories. as a seemingly transparent representation of the real body, medical photography has helped us to delve into the truth of physiognomy [Amirault 1994]. Charcot's approach was also based on the theory of physiognomy, which had a long tradition, but was thoroughly developed by Johann Kaspar Lavater in the second half of the eighteenth century. Charcot's theory was grounded in the inextricable link between man's appearance and inner state, and he referred to Lavater in many of his papers.

In his book on Lavater, Maciej Jarzewicz pointed out that one of the major problems that an art scholar or historian faces is dealing with the illustration of physiognomy, because '(...) the problem of the presence of the unconscious reading of people's physiognomy is, I think, unsolvable; you just cannot get away from this type of experience' [Jarzewicz 2013, p. 16].

Lavater was inspired by the work of Albrecht von Haller in *Elementa Physiologiae corporis humani. Tomus quintus. Externí sensus interni* (1763). He considered physiognomy to be a real scientific method, and he described it as such in his papers [Lyon 2007]. Lavater argued that human nature could be read from a face when accurately analysed [Sekula, 2010]. The facial elements he recommended subjecting to analysis were: the profile of the head and face, forehead, eyes, ears, nose, chin, etc. A man's personality was seen through the composition of all the components [Sekula 2010]. The author mentioned several hundred works of art, portraits and sketches in this publication.

Prior to Charcot, Alexander Morison worked as a doctor in Edinburgh. He diagnosed mental abnormalities, and wrote *The Physiognomy of Mental Diseases* (1843). This was not long after the invention of photography, though it had not yet entered wider usage, which is why Morison's publication reproduced the images as prints. Still, this is a perfect example of how the theory of physiognomy lies at the core of the development of psychiatric and neurological diagnostics. Morison described such basic medical categories as mania, monomania, and melancholia, basing his work entirely on how the patient's face looked, and reproducing them carefully in his publication as prints [Morison 1843]. His book became a handbook for European doctors, including Charcot [Beveridge 2004], and certainly had a large impact on nineteenth century psychiatric photography.

With the discovery of photography, there was a breakthrough in illustrating mental health, although it was still based on the theory of physiognomy, like so many nineteenth-century classification systems. The problem with the photographs of hysteria in the Salpêtrière hospital is not a general issue concerning the development of early medical photography. At the core of Charcot's work were some unique motivations, references and observations that were untranslatable into any other kind of clinical documentation.

The objectivity of photography was fundamental to its use in medicine and psychiatry. Hugh Welch Diamond was considered a pioneer of psychiatric photography. In 1848–1858 he took some remarkable photographs while director of the psychiatric hospital in Surrey County Asylum in England. He captured images which presented a few basic types of madness. In 1856, Diamond issued a document in which he described his theory of the use of photography in psychiatry. He believed that medical photography had three functions in the treatment of the mentally ill: 'it can record the appearance of the mentally ill for study (ascribing to the theories of the physiognomy of insanity accepted at that period); it can be used in the treatment of the mentally ill though the presentation of an accurate self-image; and it can record the visages of patients to facilitate identification for later readmission and treatment' [Diamond 1856]. Photographs taken by Diamond inspired Dr John Connolly, the famous Victorian Psychiatrist, to write a series of publications entitled *The Physiognomy of Insanity* (1858). The author analysed the mental state of the patient, the progress of the illness or its remission, based solely on visual material: pictures of women from the Surrey County Asylum. This proves the strength of the belief in the objectivity of visual material. As Sander L. Gilman wrote of this phenomenon: 'The Art of Truth – and representative of Truth in Art' [S. L. Gilman 1976]. This was the first time the camera became a tool for diagnosing mental illness. Charcot was familiar with Diamond's achievements and his methods, and most importantly – the functions that the Englishman attributed to photography. The Frenchman continued his work, and went a step further than his predecessor.

The photographs produced at Salpêtrière were also seen as a guarantor of authenticity. Psychiatrists and neurologists expressed their opinions on the symptoms and formulated diagnoses based on the images alone. This behaviour was a common approach to the medium in the nineteenth century. Because of the seemingly objective characteristics of the technical image, the viewer was inclined to compare the photographs to a window. He trusted them as he did his own eyes. Therefore, he criticised (and described) them not as images, but as windows to the world [Flusser 2015]. This is why the theory of physiognomy had a solid grounding in nineteenth-century photography – images were more accurate than drawings, and thus they conveyed the idea of facial features.

Though I use the term 'Charcot's photographs', this is not entirely accurate. He did not take the pictures themselves, he only had them commissioned. He worked primarily with two photographers: Albert Londe and Paul Richter. Charcot, however, called himself a photographer and an artist and considered himself to have made those illustrations, as he determined the appearance of every patient and scene, and as such, the physiognomy of hysteria.

It was Regnard, not Londe, as is commonly believed, who took the famous photographs of hysterics Augustine and Blanche Wittman. But Albert Londe became the most famous photographer at Salpêtrière. Though he was not a doctor, in 1883 he became the director of the photography department. He founded a photography studio in the hospital. He was the first professional photographer in Europe who was employed full-time at a medical institution [Gilman et al. 1993]. He was one of the greatest innovators of psychiatric photography. Londe's works were published in every edition of *Nouvelle Iconographie* until 1918, becoming the most impressive collection of photographs of the psychiatric *fin-de-siècle*.

Since photography was very important to the work of the Salpêtrière hospital, a specialised, well-equipped studio was established. Photographers worked with modern photographic equipment, using a multiple-lens method. The technology evolved with the increase in knowledge. Still, the process of taking a single photo was time-consuming and required long-term exposure (an average of fifteen minutes). This raises the question of how such dynamic poses could have been captured. The only answer is that the women were in fact posed. It was necessary for them to remain motionless, static and still.

The medical subjects in these photographs were posed. Paradoxically, these fabricated images passed as objective information and confirmed the patient's pathology. The patient existed only in relation to the illness she performed, as directed by the doctor. From this point of view, medicine and photography are identical [Amirault 1994]. And although Charcot firmly established his theory in the belief that he was creating reliable photographic documentation and communicating the truth, the whole act of creating a photo was based on women performing the convulsions required of them. Photography was an autopsy performed on living hysterics [Didi-Huberman 2003]. Assuming the authenticity of the existence of the referent, however, led to a paradox where hysteria was concerned: it authenticated existence, but through theatrical means.

The collapse of faith in the objectivity of photography that occurred in the second half of the nineteenth century engendered a collapse of faith in hysteria.

Hysteria had a face and this made it the truth. It appeared in a number of scientific publications; psychiatrists and neurologists analysed it and drew their own conclusions, based on no more than a photograph. A photograph of a face was considered to be a document, and physiognomy a reliable material for analysis. I would like to describe a few examples of these photographs.

The first example is Chart II from *Iconographie photographique de la Salpêtrière* (first edition, 1876–1877). The photo was taken by Paul Regnard, and depicts the first phase of a hysterical attack: the epileptic phase.

The woman is shown lying on the bed, the photo is taken from the perspective of the feet. Only her face is visible from under the blankets and quilts; it seems to be calm, but her hair is untidy and her outline is blurred due to an overlong shutter-speed during dynamic movement. The woman is on the edge of the bed, her hands are twisted, her eyes closed and mouth slightly ajar, the top of her face is the only fully visible part of her body, and so the physiognomy provides an image of hysteria. This phase was characterised by an attack that precisely resembled epileptic seizures. The patient lost consciousness, completely succumbing to convulsions [Richer 1881]. Hysterical symptoms were thought to be borrowed from others the patient observed in the hospital [Bogousslavsky, Ettlingen 2014]. The straight lines of the bed are arranged to contrast with the chaotic folds of the quilt. The photographer is far from the model's face, but he uses an interesting setting.

The second example is the twelfth plate from *Iconographie photographique de la Salpêtrière* (second edition, 1878). The photograph was taken by Regnard and shows the attack in the tonic phase. The patient is shown again on the bed, from the waist up, but this time from a certain distance, in profile. The woman is reclining, a position suggested by the shape of the bed's metal frame. Her face expresses ignorance, she has tightly clenched lips and eyes, tousled hair, her chin is tucked in to her chest. The hands of the patient are unnaturally twisted outward. The tonic phase was characterised by muscle tension throughout the body, lasting from a few seconds to to several minutes [Charcot, Richer 1887].

Charcot's collection of physiognomic images adhered to the trend for photographic archives. An archive is a paradigmatic whole; between 1880 and 1910 it became an essential tool in empirical disciplines. Alan Sekula has pointed out that archival rationalisation was most necessary for modes of photographic realism that were instrumental, which were designed to contribute directly (or indirectly) to the practical conversion or manipulation of their referent [Sekula 2010]. Researchers point out that, with the popularisation of photography, the inventory, archive and symbolic subordination began to arise from the imperative to record the surrounding reality which was common in emerging industrial societies [Rouillé 2003].

The basic method of nineteenth-century science was comparing photographs of the body. Jarzewicz wrote: 'The comparative analysis of the history of art and culture has a long tradition in theory and practice; its chief criticism has been its failure to meet hopes for gaining new knowledge about the test subject (…). Although everything seems to be comparing images as a method of capturing, systematising and transferring knowledge, it remains one of the most important facets of intellectual activities of any kind' [Jarzewicz 2013]. In the nineteenth century, physiognomy became a basic methodology for imagining and diagnosing mental illness. As a seemingly transparent representation of the real body, medical photography has helped us obtain evidence of the truth of physiognomy.

Charcot's archive was created parallel to the growth in popularity of nineteenth-century forensic systems. In France, the pioneers in this field were Alphonse Bertillon and Francis Galton; similar systems were also founded in other countries, e.g. in Italy by Cesare Lombroso [Sekula 2010]. Charcot was certainly familiar with their practices and drew inspiration from their methods. They required a precise and rigorous adaptation of forms and images, aesthetic issues inextricably tied to administrative, organisational and technical aspects. These phenomena must have influenced the neurologist in his own work.

To emphasise the important role the face played in the creation of typologies, we ought to mention the electric shock experiments. Duchenne de Boulogne, also a doctor in Salpêtrière, worked on the subject before Charcot. He is seen as Charcot's master and teacher. The photographs he commissioned depict the moment when the patient's face was contorted by sharp electric shocks. The body became twisted. This was a way of presenting the work of the facial muscles. But the truth is also ambivalent, as actors were often employed [Trumpler 1993]. Although these photographs became popular, we also know that Charcot conducted similar experiments and documented them, inspired directly by Duchenne de Boulogne.

We might ask: what was the truth? What do we see in these photographs? The question remains unanswered, and we only have hypotheses on what we witness in these images and what took place in the Salpetriere hospital.

One answer might lie in Charcot's experimental therapies. When hypnosis was discovered, it immediately caught the neurologist's attention, and he quickly implemented it at the Salpetriere hospital. He developed a method of using hypnosis in treatment. During sessions he held each Thursday for groups of interested Parisians, he demonstrated hysterical attacks and other abnormalities by putting women under hypnosis. In this state they could pretend to be animals, eat chalk and have highly erotic convulsions [Nicholas, Gounden, Levine 2011]. The neurologist achieved the desired hysterical appearance through hypnosis and suggestion. As Jonathan Crary wrote: 'in the [18]80s discussions were held in which they deliberated whether this enigmatic condition was, as believed by J.-M. Charcot and his successors at the Salpêtrière, a somatic symptom of an underlying disease, or an exaggeration, a completely normal reaction to suggestion' [Crary 2009].

Charcot conducted other experimental therapies which served for visualisation and photographic documentation. Women were exposed to extreme temperatures and inhalations and were drugged. Sometimes the hysterics died from exposure to ether, alcohol, morphine and other toxic substances [Didi-Huberman 2003].

Some believe that the women decided to act like hysterics on purpose. Didi-Huberman postulates this theory in part, as does Dorota Sajewska in her book *Chore Sztuki* [Sick Arts]. In a patriarchal society, marginalised through poverty, deformation or other reasons, women could reconstitute themselves by being hysterics, drawing the attention of society and having their place in this exclusion. Hysteria is often described as a women's disease, and it is they who appear in most of the pictures from Salpêtrière. This is why the female aspect in this phenomena is accentuated.

The face played an important role throughout the process – it performed the changing states of mind, moving from one phase of an attack to another; it also showed unconsciousness and distance from reality. Nevertheless, hysterics' physiognomies were exaggerated.

Either to meet the demands of a psychiatrist, or under hypnosis or drugs, women fell into convulsions. We can only speculate to what extent patients were aware that their bodies were a 'lie'. Variability in the appearance of symptoms was also problematic for the scientists of the *fin-de-siècle*. Charcot perceived this problem, but found an explanation, by which the symptoms were true despite the duplicity of the patient. 'We meet with [simulation] at every step in the history of hysteria, we can admire the incredible art, wisdom and perseverance with which women, under the influence of the great neurosis, break the rules of the game – especially when a doctor is a victim' [J.-M. Charcot 1881, p. 264]. Such an approach, paradoxically, did not negate the disease, but only accounted for its characteristics. For a doctor, it was more difficult to give a correct diagnosis and properly recognise the ailment. Dubois wrote: 'the hysteric is an actress, a *comedienne*; but never say this to her, because she does not acknowledge the game, she sincerely believes in the reality of the situation' [P. C. Dubois 1904, p. 214].

Charcot's hysteria appeared in a transition period, in a liminal moment for both knowledge of mental illness and for the development of photography: a few decades after the invention of photography, but before the flood of images; more than half a century after medicine first took an interest in mental illness, but before precise, scientific descriptions emerged. Charcot's era was the only time when a combination of photography, psychiatry and the human imagination could produce such fascinating results.

Hysteria was created through physiognomy, and more specifically, through the manipulation of physiognomic appearances. The intervention of the creator, Charcot, was omitted, and belief in his theory became the prevailing view for almost twenty years. The result is a typology of what never existed, but paradoxically, because of its lack of referent, it has become an important moment in the history of culture and science.

Bibliography

1. Amirault, C., 1994. 'Posing the Subject of Early Medical Photography', *Discourse: Journal for Theoretical Studies in Media and Culture*, No. 2.
2. Beveridge, A., 2004. 'Psychiatry in pictures', *British Journal of Psychiatry*.
3. Bogousslavsky, J., Boller F., 2013. 'Jean-Martin Charcot and art: Relationship of the 'founder of neurology' with various aspects of art', *Progress in Brain Research*.
4. Charcot, J.-M., Bourneville, D.-M., Babinski, J. F. F., 1892. *Oeuvres completes de J. M. Charcot. Leçons sur les Maladies du système nerveux*, Vol. 1, Bureaux du Progrès Médical, Paris.
5. Charcot, J.-M., 1881. *Lectures on the Disease of the Nervous System*, New Sydenham Society, London.
6. *Historia ciała*, 2013. Vol. 2, ed. A. Corbin, trans. K. Belaid, T. Stróżyński, Gdańsk.
7. Didi-Huberman, G., 2013. *Invention of Hysteria. Charcot and the Photographic Iconography of Salpêtrière,* trans. A. Hartz, The MIT Press, Massachusetts.
8. Foucault, M., 1987. *Historia szaleństwa w dobie klasycyzmu*, trans. H. Kęszycka, Państwowy Instytut Wydawniczy, Warsaw.
9. Gilman, S. L., King, H., Porter, R., Rousseau, G. S., Showalter, E., 1993. *Hysteria Beyond Freud*, University of California Press, Los Angeles.
10. Gilman, S. L., 2014. *Seeing the Insane. A Visual and Cultural History of Our Attitudes Toward the Mentally Ill*, Echo Point Books & Media, LLC, Vermont.
11. Goetz, S. L., Bonduelle, M., 1995. *Charcot: Constructing Neurology*, Oxford University Press.
12. Jakubik, A., 1979. *Histeria: metodologia, teoria, psychopatologia*, Państwowy Zakład Wydawnictw Lekarskich, Warsaw.
13. Jarzewicz, M., 2013. *Sztuka i wizualizacja naukowa: ilustracje do Fragmentów Fizjonomicznych Johanna Caspara Lavatera*, Neriton, Warszawa.
14. Lyon, J. B., 2007. 'The Science of Sciences: Replication and Reproduction in Lavater's Physiognomics', '*Eighteenth-Century Studies*', No. 2.
15. Morison, A., 1843. *The Physiognomy of Mental Diseases*, Longman, London.
16. Nicholas, S., Gounden, Y. and Levine, Z., 2011. 'The Memory of Two Great Mental Calculators: Charcot and Binet's Neglected 1893 Experiments', *The American Journal of Psychology*, No. 2.
17. Richer, P., 1881. *Études cliniques sur l'hystéro-épilepsie ou grande hystérie*, Bureaux du Progrès Médical, Paris.
18. Rouillé, A., 2007. *Fotografia. Między dokumentem a sztuką współczesną*, trans. O. Hedemann, Univeristas, Krakow.

19. Scull, A., 2009. *Hysteria. The disturbing history*, Oxford University Press, New York.
20. Sekula, A., 1986. 'The Body and the Archive', October, Vol. 39 (Winter, 1986), pp. 3–64.
21. Shorter, E., 2005. *Historia psychiatrii: od zakładu dla obłąkanych po erę Prozacu*, trans. P. Turski, Wydawnictwa Szkolne i Pedagogiczne, Warsaw.
22. Sontag, S., 1977. *On Photography*, Farrar, Straus and Giroux, New York.
23. *The Face of Madness: Hugh W. Diamond and the Origin of Psychiatric Photography*, 1976. trans. S. L. Gilman, Brunner/Mazel Publishers, New York.
24. Trillat, E., 1993. *Historia histerii*, trans. Z. Podgórska-Klawe, E. Jamrozik, Ossolineum, Wrocław.
25. Trumpler, M., 1993. 'Reviewed Work: The Mechanism of Human Facial Expression by G.-B. Duchenne de Boulogne', *Isis*, No. 1.

Ewelina Drzewiecka

The significance of face in John of Głogów's *Phisionomia hicinde ex illustris scriptoribus* (1518)

The subject of this paper is John's of Głogów Latin dissertation called *Phisionomia* (1518). This work – a study of man from medical, psychological and physiological perspectives – reflects the belief that character lies in the body's physical attributes. This article presents a brief overview of the treatise's content, indicating the aspects that the author finds most important: the face and its significance.

Keywords: John of Glogow, *Phisionomia*, physiognomy, face, Renaissance treatise

In facie legitur homo.

(*A man's character lies in his face*)

Known and practised for centuries, physiognomy was a meeting point of many other disciplines, such as psychology, medicine, philosophy and astrology. It stemmed from the perennial desire to understand human nature, and the belief that its secrets could be learned by observing the body. There have been many attempts to define physiognomy; the simplest definition is that physiognomy is the reading of a man's character from the physical features of his body and, most importantly, his face. Although the scientific legitimacy of physiognomy has been often called into question, it has aroused keen interest for centuries, and many renowned medical doctors and philosophers have appreciated and used its potential, to name only Hippocrates, Galen of Pergamon, Aristotle, Giacomo della Porta or Girolamo Cardano. Another proponent of the discipline was John of Głogów, or Johannes Glogoviensis (ca. 1445–1507), a polymath and a professor of the Jagiellonian University, who, in his forty years of academic activity, helped to educate many of the brightest minds of his epoch (probably including Nicolaus Copernicus). His interests were many and diverse, encompassing most of the subjects taught at the university, and he left behind a host of writings on philosophy, astronomy, astrology, grammar, optics, medicine and geography. Some of John of Głogów's works have been preserved in manuscripts; of others, there are extant print versions. Many of these works are compilations [Birkenmajer 1937, p. 24], although interspersed with new and original thoughts [Zwiercan 1963, p. 451]. The vision of anthropology that emerges from his works is the study of man from philosophical, medical, psychological and physiological perspectives. The subject of my paper will be John of Głogów's Latin dissertation called *Phisionomia*. This interesting book, printed in the shot quarto format in 1518 in Krakow by Rudolfus Agricola Iunior, has yet to be translated from the Latin. Although a few quotations from this work have been published in other books [Domański 1978, pp. 439–443; Chojecka 1969, pp. 177–180] and articles [Dębicki 2008, pp. 97–100], it has never been the subject of in-depth study. It is worth noting that Glogoviensis's book was the only physiognomical treatise published in Central and Eastern Europe during this period [Porter 2005, pp. 80–81]. As such, my paper is only a prelude to further research, to shed some light on parts of this forgotten and fascinating work. The majority of this paper will contain Latin quotations of some untranslated parts of *Phisionomia* with English translations. It will show clearly how the significance of the face is presented in this work.

In ancient times, there were three fields that dealt with physiognomy: medicine, philosophy and psychology [Sosnowski 2010, p. 9]. Like many other disciplines of ancient learning, physiognomy was rediscovered in the Middle Ages via Arabic translations that had preserved and built upon the findings of Greek philosophers [*cf.* Ghersetti, Swain 2007, pp. 309–329]. In addition to the naturalist approach, the discipline also began to be connected with astrology at the time. This resulted in the publication of numerous physiognomic and astrological treatises on aspects of chiromancy and oneiromancy. There were also plenty of treatises written by doctors of medicine, which usually contained a list of parts of the body, the four humours and temperaments [Coutrine, Haroche 2007, p. 34]. The sixteenth century ushered in a few innovations in the field of physiognomy. The first was the rediscovery of the ancient relationship between soul and body: the body was subjected to more detailed scrutiny, as some believed that by studying its 'semiology' one could gain insight into the 'natural proclivities of the soul'. Another change in the early modern period was the reappreciation of Aristotle's hierarchy of body parts, with the primacy of the head and, by extension, the face. Finally, the study of man from the broader perspective of his mutual interaction with the natural world prompted a return to the findings of Pseudo-Aristotle, which were based on analogies between human and animal features. It could be said that early modern society prefigured this problem. In this period there were plenty of chronicles that described monstrous beings incorporating human and animal components, so *nil miradum,* that people 'were also well habituated to the less extreme and rejudical ideas, involving the crossing of the human-animal frontier. One of these was physiognomy' [Jones 2012, p.9, cf.: Davidson 1991; Daston, Park 1998].

In light of these facts, John of Głogów's treatise, published in 1518, straddles the boundary between Medieval and Renaissance physiognomy studies. The title of the treatise: *Phisonomia hincinde ex illustribus scritptoribus per venerabilem virum magistrum Joannem Glogoviensem diligentissime recollecta.* ('Physiognomy, or [an excerpt] from [the writings of] renowned men of letters, compiled by the esteemed *magister* Johannes Glogoviensis') suggests that the author appeals to *auctoritas*, to the Ancient and Medieval authorities. And though the book is still Medieval in many respects, there are clear signs of tendencies that were to become characteristic for later treatises in the field.

As a philosopher and medical doctor with a deep interest in the study of nature, John of Głogów often cites Aristotle and his naturalist psychology [Kuksewicz 1963, p. 183]. References to astrology, on the other hand, are few and far between in *Phisionomia*. This may be somewhat surprising, given that Glogoviensis' academic work coincided with the heyday of the Krakow School of Astronomy and Astrology, a leading university in the field in Central Europe. This made Krakow an internationally renowned centre of celestial studies, attracting many scholars from faraway countries [*cf.*: Birkenmajer 1956]. Evidence of this may be found in many chronicles of the time and later, which reported that *Cracovia astrologis referta est* – 'Krakow is overflowing with astrologists' [*Historia critica regnum Hungariae* 1793, p. 258]. It is worth noting that a shared department of mathematics and astronomy was created at the Jagiellonian University in 1410, one of many such departments throughout Central Europe. The great strides in astronomy were not matched by the development of astrology. However, in 1459, Marcin Król of Żurawica founded a department of astrology at the Jagiellonian University that was unparalleled in Central Europe. At this time, only a few Universities in Italy had comparable departments. Owing to these departments, Krakow became an international astronomical and astrological centre, and an outstanding astrological school in Central Europe. Its high university standards and prestige are shown in that Jagiellonian graduates became professors at foreign universities or found employment at royal and papal courts throughout Europe. In the last quarter of fifteenth century and in the first decades of sixteenth century, through the graduates who stayed in Krakow and became professors there, the Jagiellonian University began attracting many foreign scholars. Among them was John of Głogów [Birkenmajer 1956, p. 366].

In light of contemporary research, John of Głogów's astrological writings can be divided into three groups: the first and largest might include prognostics and astrological calendars which, because of their ephemeral and temporary nature, remain preserved in fragments. The second group contains textbooks that describe how to read and use the astrological calendars. The most interesting writings of the *Glogoviensis* can be found in the third group. Those are treatises on change in the weather (*Opusculum de mutatione aëris),* horoscopes (*Introductorium in scentiam navitatium*), choosing good days (*Tractatus de electionibus seu tempora electa*), the defence of astrology (*Defensio astrologiae*), geography and astronomy (*Introductorium cosmografiae*) and, finally, the very interesting *Tractatus de 48 imaginibus caelestis,* which describes celestial constellations in light of astrological and mythological interpretation. As the author of many astrological treatises, John of Głogów enhanced the school's prestige. He also gained international renown by writing horoscopes and prognoses. Even so, his *Phisionomia* shows a marked deviation from astrology and signals a turn that was to become even more pronounced much later, in works published in the 1580s, and especially in the treatise by Giovanni della Porta, in which the Aristotelian naturalist tradition eventually supplanted the old astrological one. In the foreword to his book, Glogoviensis supplies a definition of physiognomy that relies on the authority of Aristotle and Constantine the African: he defines it as 'the skill and knowledge that enables one to identify the body's natural proclivities through observing its external parts' (A2). The following is a brief overview of the content of this treatise, indicating the issues that its author considered most important. Part One is devoted to all the parts of the human body, with a description of their characteristic features. The author starts from the head and face and then describes the eyes, forehead, cheeks, chin, ears, nose, mouth, lips, tongue and teeth. Then he moves on to the larynx, artery, throat, neck, nape of the neck, shoulders, arms, hands, fingers, nails, then the middle and lower parts. John of Głogów wanted to see the human body in its entirety, he moves from the whole to the part, and thus his descriptions are sometimes very detailed: after the skin, he lists the pores, flesh, nerves, and finally, the bones. Part Two concerns the general division of the human body. Drawing on the authority of Constantine the African, the author divides the body into three sections: upper, middle and lower. John of Głogów considers the upper section, which contains the head and face, to be the most important. Part Three, entitled *Physiognomy,* describes the head and its features. Part Four, entitled *De colore,* focuses primarily on the face, analysing the skin tone and its relationship to character and habits. Part Five correlates hair volume, type and colour with character and proclivities. Part Six concentrates on the eyes and how character traits can be correlated with their position, shape and size. Part Seven focuses mostly on the face, forehead and lips. Part Eight describes the nostrils and nose with reference to Aristotle and other philosophers. Part Nine analyses the other body parts, such as lips, ears, neck, teeth and tongue. Part Ten covers the chin and beard – including their shapes and the impact they have on one's habits. Part Eleven is devoted to such parts of the body as the nails, shoulders and ribcage. Part Twelve describes the spine, stomach, ribs and legs. Part Thirteen, the final chapter, focuses on the parts of human body which John of Głogów failed to mention. This cursory overview reflects the structure of the content in *Phisionomia*, which closely follows the structure found in medieval medical tracts. The author puts special emphasis on the upper body, particularly the face. I will now proceed to present passages devoted to the face and to point out several traditions observed therein. The Aristotelian tradition prevails in the suggestions that the external physiognomic features reflect a person's character and betray the characteristics of the 'inner man'. The first quotations will concern the face:

Part VII, 1: A man whose face is very fleshy and pale is idle and drowsy; he does not ponder the things he does; he is indiscreet and has a bad memory. A man whose face is petite and too slender is envious; he badmouths people [...].

Differentia VII, 1: Cuius facies multum carnosa est et alba homo est piger, somnolentus/otiosus, de rebus agendis parum cogitans, discretionis impeditae et malae memoriae. Cuius facies et nimis extenuata et macilenta hominem signat invidosum, de aliis male loquentem [...]

Part VII, 3: A man whose face is very round is foolish. An uneven, non-symmetrical face is the sign of a man with bad habits, who is suspicious in his malignant heart. He is a disparager of people. A face that is too oblong signifies a foolish and unpleasant man with a rotten heart. A slightly oblong face is a sign of prudence and resourcefulness [...]

Differentia VII, 3: Cuius facies vehementer rotunda est, stolidus est. Cuius facies difformis est malorum est morum, cordis maligni suspicious, hominum deturpator et detractor. Cuius facies nimis oblonga fatuus est et inuerecudus et cordis mali. Facies in parte oblonga prudentiam et astutiam significat [...].

Part VII, 5: A face that is quick to sweat signifies a man of a warm nature who eats sumptuously and vomits. He is a man of sharp intellect but also vain and shameless, indiscreet. As Constantine maintains, he whose face is very oily and wide is a simpleton of little discernment. His intellect is blunt and he is forgetful [...] He is a scoffer, vain at heart, who disparages others. He is more evil than good.

Differentia VII, 5: Facies hominis quae cito sudat signat hominem calidae naturae, multa comedentem et vorantem. Est homo duri ingenii, vani cordis, impudicus, non bonae discretionis. Cui facies est excallenter pinguis, inquit Constantinus, et lata hominem signat semplicem, pauca prudentia in rerum cognitione [...]. *Est hominem grossi ingenii et intellectus. Est obliviosus, vani cordis, est detractor aliorum derisor. Plus malus quam bonus.*

Part VII, 6: People whose faces are flat and not bloated are prudent and friendly, they are righteous, and their conversation is proper. As Constantine concludes, a man with a timid face is wont to drink, he eats a lot, his intellect is meagre, and he is indiscreet. He is impudent and foolish, conceited and condescending, and confuses other people. Conversation with him is distasteful and elusive. He often lies and is garrulous; he makes grand promises but hardly ever delivers.

Differentia VII, 6: Cuius facies est quasi plana et sine tumore, est homo prudens, discretus est signum prudentiae et amicitiae, est legalis animi et rectae conversationis. Concludit Constantinus cuius facies est tumida, est bibulus, homo ebriosus, multa comedens, pauci intellectus et dicreditionis, inverecundus, stolius, multa de se confidit, alios confundit et se ipsum extollit. Eius conversatio est odiosa et fugienda. Est homo qui multa mentitur et loquitur. Multa promittit paucam facit.

The next example concerns the forehead:

Part VII, 8: People whose foreheads are flat and wrinkle-free are quarrelsome and prone to provoking arguments with others; their hearts and souls are full of malice. As Constantine maintains, people with large foreheads are idle, indiscreet and of meagre intellect; they are stubborn. Those with small foreheads are foolish and slow-witted. A man whose forehead is full of wrinkles is shameless, undisciplined and malicious, he disparages others and says bad things about them, he is a two-faced scoffer: he says one thing and thinks another.

Differentia VII, 8: Cuius frons est plana, rugas non habens, homo est litigiosus, rixosus et alios in rixas ducens. Est homo maligni cordis et animi. Inquit Constantinus, cuius frons magna est, piger est et parvi intellectus et discretionis. Est homo proprii animi et pertinax in rebus. Cuius frons parva multum, stolidus est et duri ingenii. Cuius frons multum rugosa inverecundus est, indisciplinatus, malus, detractor aliorum, irrisor et mala de hominibus loquens. Est homo duplicis animi, aliud in ore, aliud in core habet.

Part VII, 9: A man with a round forehead is bountiful, has a considerable intellect, and is discreet; he is likeable and liked well enough, he has a joyous disposition, as Constantine says. A very fleshy forehead signifies indiscretion and impudence. It also signifies a man of proud soul and vain heart [...]

Differentia VII, 9: Frons in rotundum disposita hominem legalem et largum designat. Est homo discretionis bonae et facilis intallectus. Est homo qui favorem hominum habet. Est hilaris animi, in omnes leti est animi et tristitias non admittit. Inquit Constantinus. Frons multum carnosa signat inepritudinem ingenii et paucam discretionem; signat hominem superbi animi et vani cordis [...].

John of Głogów's *Phisionomia* also contains a great many remarks about the eyes:

Part VI, 1: Rasis Almansor says that by analysing the position, size, colour and features of the eyes and their depth, one can learn a man's habits, both good and bad. Hence, those who were trained in this art wrote a great deal on the significance of the eyes and saw that the course of a man's life depended on their placement. A man whose eyes are large is [by nature] idle and careless; he cares not for order and relies on Fortune completely. A man whose eyes are large and unfocused is idle and drowsy, an unpleasant fellow. He likes to woo women and his faith is weak. His speech is [apparently] flippant and his heart is full of iniquity. He speaks what he does not think.

Differentia VI, 1: Inquit Rasis Almansor, ex oculorum dispositione, colore, quantitate, eminentia et profunditate eorum in capite maxime hominis dispositio in moribus suis agnoscitur, de bono et de malo. Unde et huius artis magistri distincte de colororum significationibus scripserunt et pro maiori parte totam hominis vitam dispositioni oculorum attribuerunt. Cuius enim oculi magni sunt piger est, incurabilis, rerum suarum ordinem non habens, omnia fortunae comittit. Cuius oculi magni sunt et tremuli homo piger est et somnolentus, inverecundus, mulierum amator. Est homo cui fides est pauca. In euius sermone instabilitas. Cor non habet rectum. Quod ore loquitur corde non intendit.

Part VI, 2: A man whose eyes bulge is ill-mannered and garrulous and foolish; he believes that all he does is prudent. A man with deep-set eyes sees well and distinctly; his judgement is sound and his memory prodigious. He is shrewd, discreet and resourceful. As Constantine says, a man whose eyes are small and unfocused is ill-mannered and vain of heart; he leads people astray and fosters evil.

Differentia VI, 2: Cuius oculi multum foris eminent, inverecundus est, loquax et stolidus. Omnia prudenter agere credit. Cuius oculi in profundo positi sunt, boni et acuti est visus, rerum considerationem fortem habet et discretionis, memoriae magnae et bonae. Astutus est, discretus et callidus. Cuius oculi parvi inquit Constantinus et treumuli, inverecundus est, vani cordis, aliorum seductor, et malorum causa et origo.

Part VI, 3: A man whose gaze flickers as rapidly as [a flame] in a lamp is cunning and resourceful; he does his business judiciously and seeks his own gain; his heart, however, is vain and malicious. It is no avail to speak with him. His heart and soul are full of malice. He is duplicitous and says not what he thinks. A man with enormous eyeballs is timid; he sees well, but not from afar. Eyes like a goat's are healthy and of sound sight; they rarely fail. Even so, a man with such eyes is dim-witted, shameless and cunning. He demeans other people.

Differentia VI, 3: Cuius oculi cito moventur ut in lampade, astutus est et callidus, res suas prudneter agit et suam curat utilitatem, homo tamen vani cordis et mali. Est homo cui conversatio hominibus non prodest, plus enim maligni est cordis et animi quam recti. Est homo duplex, loquit unum ore et aliud in corde habet. Cuius oculi multa magnitudine apparent, timidus est, bene videt sed non a remoris. Cuius occuli assimulantur oculis caprae, sani sunt et boni visus, raro infirmantur et debilitantur, homo tamen talis stolidus, inverecundus est et callidus. Est homo aliis detrahens et vitam aliorum reprehendens.

Part VI, 11: Small and round eyes betray a generous, simple, pleasant man; such a man will believe what people say, and is thus gullible. He is neither sharp-witted nor very shrewd. He is quick to anger and his ire lingers. As Constantine maintains, a man with slanted eyes is prudent, but also crafty. He is quick to anger and envious. He desires more than he yields; he is doubly fickle and is more mendacious than veracious. He has a bad disposition; his discourse is unpleasant and flippant.

Differentia VI, 11: Oculi parvi et ad rotundum inclinati hominem liberalem ostendunt, simplicem tamen et verecundum, cito his quae dicunt credentem, unde sepius fallitur. In ingenio est difficlis et in ingenio est gravis, cui commovetur ad iram vindictam appertit eam diu tenet. Inquit Constantinus, cuius oculi ad obliquitatem tendunt, homo est prudens sed plus callidus, iracundus, invidus, tenacis manus. Plura cupit accipere quam dare, est inconstans plura mendit quam vera dicit, maligni est animi et dolosi eius conversatio inter homines est ingrata et fugienda

John of Głogów also writes about the nostrils:

Part VIII, 1: As Constantine says, those enamoured of literature have delicate nostrils; they are discreet and intelligent but also cunning and shrewd, and prone to mockery […]. Large nostrils are the sign of a fool.

Differentia VIII,1: Inquit Constantinus cuius nares subtiles, est homo dilignens litteraturae, boni ingeni et discretionis sed est astutus, callidus, alios homines circumveniens […]. *Cuius nares grosse sunt est homo stolidus, paucae sapientiae* […].

Part VIII, 10: A man whose nose tends to be ruddy and warm is avaricious and stubborn. His skin will scar and discolour easily. He is duplicitous […]

VIII, 10: Cuius nasus ad ruborem vergit complexionis est calidae, homo avarus et tenax. Est homo declinatus ad infirmitates graves ad morpheam et lepram est homo duplicis animi […].

Another tradition found in part in John of Głogów's work is the description of human appearance through animal analogies. As before, this tradition also stems from *Physiognomy* by (Pseudo-)Aristotle.

Part VII, 7: In his *Physiognomy*, Aristotle maintains that people with fleshy faces are prone to lusting and are easily entangled in lechery – they behave like oxen. Those with emaciated faces are restless and timid, like deer and asses. […] Aristotle concludes that it is best to have a face that is neither too large nor too small.

Differentia VII, 7: Inquit Aristoteles in Phisionomia sua qui faciem carnosam habent facile concubiscibiles sunt et ad vitae impudiciam inclinati et refuerentur ad boves. Qui faciem macilentiam habent solliciti sunt et timidi et referuntur ad cervos et asinos. […] *Concludit Aristoteles qui neque parvam neque magnam oportet esse* […].

Part VI, 8: A man whose eyes are large, like an ox or a bull, is dim-witted, of little discernment and little intellect; he believes anything that is passed as truth. His memory is bad, he acts impetuously and lacks manners. Dark eyes of moderate size signify a man of a sound intellect, refined manners and eloquent discourse. Such a man is righteous; he loves peace and harmony and can coexist with people.

Differentia VI, 8: Cuius oculi sunt valde grossi similitudinem habentes cui oculis vituli aut bovis, est homo mali intellectus, stolidus, pauca discernens; quidque ei dicitur verum credit. Est homo malae memoriae, homo simplex et stolidus [...], *brutaliter omnia agit, nec est urbanitas in eo. Cuius oculi sunt mediocres in quantitate ad nigredinem tendentes, hominem signant legalem,boni intellectus, benignae et humanae conversationis. Est homo legalis cum hominibus componens, pacem dilignes et concordiam.*

Part VII, 10: Following Aristotle: people with small faces are undisciplined and resemble pigs. Those with very large faces are slow and resemble oxen. Round and fleshy faces mark idle men, who resemble asses. People who are envious have overly angular faces; they resemble dogs. Quasi-square faces signify the impudent; these resemble lions. People with sad features are melancholic. Those with wide and drawn out faces are flatterers who mock others.

Differentia VII, 10: Qui frontem parvam habnet indisciplinabiles sunt et referuntur ad sues. Qui valde magnam, referuntur ad boves et tardi sunt. Qui rotundam excellenter et carnosam pigri sunt et referuntur ad asinos. Qui nimis longam invidi sunt et referuntur ad canes. Qui quasi quadratam audaces sunt et referuntur ad leones. Qui tristem frontem habent melancolici sunt. Qui frontem habent extensam et latam adulatores et aliorum irrisores sunt. Haec Aristoteles.

Part V, 3: As Aristotle maintains: delicate hair is a sign of a timid person while strong and coarse hair marks the courageous. The philosopher came to this [conclusion] by observing animals. The most skittish animals are the deer, the hare and the sheep, whose hair is very delicate. The lion and the boar, on the other hand, have coarse fur. This holds true for birds, too: those with rough feathers are strong, like the swan or the duck, while those whose feathers are soft and delicate, like the dove, are timid by nature. Aristotle also contends that those who live in the north have coarse hair but are not among the brightest. Those who live in the south have softer hair and are wiser.

Differentia V, 3: Inquit Aristoteles, pili molles timidum signant, duri fortitudinem. Hoc enim signum sumptum est ex animalibus, timidissimis igitur cervus. lepus et ovis pilum mollissimum. Habent fortitudinem autem leo et aper et pilum durissimum naturaliter. Apparet et hoc in avibus quae enim duram habent pennam fortiores sunt, ut cignus et anser, quae vero molliores timidiores ut columba. Subiungit Aristoteles, qui aput septentironem habitant fotres sunt et durorum pilorum et minus sapientes. Qui vero ad meridiem timidi et pilum mollem habent et boni intellectus.

Part VIII, 7,8: A large nose is a feature of a shameless man, while a pointed nose marks those who are talented, intelligent and cunning. Such people do not speak their mind, are envious and self-made. According to Aristotle, they resemble dogs.

Differentia VIII, 7,8: Nasus magnus impudicitiam signat [...]. *Acutus nasus hominem ingeniosum signat boni ingenii et intelectus, est callidus et astutus* [...] *in verbis suis bene loquitur et in corde aliud cogitat. Est homo invidus* [...]. *Secundum Aristotelem assimulatur cani.*

Based on the passages excerpted above, it might be observed that John of Głogów uses two methods of deduction sourced from the (Pseudo-)Aristotle's treatise: (1) direct deduction, whereby individual physiognomic features refer to specific psychological ones (e. g. people with large noses are lecherous) and (2) indirect deduction, whereby a psychological feature is deduced by drawing analogies, especially from the animal kingdom (a lion's muzzle is square and lions are courageous and impetuous, therefore people with square faces must also be courageous and impetuous). These two methods are used interchangeably throughout the treatise. Additionally, quasi-medical references may be found in the work. As a medical doctor, John of Głogów notes that some external features of the face may be signs of various disorders or malfunctions of the internal organs. This combination of health-related attributes with characterological studies yields very interesting results.

Part VII, 4: A face that is too wan by nature is the sign of a sickly person who is prone to emaciation. This indicates bad digestion and a frail liver. Such a man is envious, cunning, avaricious, untrustworthy, prideful and vain of heart. A ruddy glow on the skin marks a man who is good, faithful, discreet and wise [...].

Differentia VII, 4: Cuius facies a natura est multum pallida hominem insanum signat et de facili in aegritudines inclinatum. Signat malam digestivam et epatis infirmitates, est talis invidus, tenax, avarus in rebus fidem non tenet, est superbus et vani cordis. Cuius autem facies in colore vivida, hominem boni animi ostendit, fidelem, bonae discretionis et sapientem [...].

Part VII, 11: A man with plump lips is foolish, shameless and prone to mocking people. Such man is none too wise and indiscreet; his speech is disjointed and he talks of foul things. As Constantine believes, a man with pale lips is weak and has a failing liver. This is evident in his frailty: the blood that should circulate through the extremities is concentrated in the internal organs. Pleasantly-coloured lips are a sign of health and good metabolism.

Differentia VII, 11: Cuius labia magna sunt, stolidus est, inverecundus, aliorum irrisor. homo paucae sapientiae et discretionis, multa loquit et immoderate et turpia. Cuius labia, inquit Constantinus, non bene colorata sunt debilitatem habet interioribus et debilitatem et defectum in epate. Huius rei apparentia et probatio est in his qui in infirmitatibus laborant, sanguis enim qui est in exterioribus corporis partibus, ad interiora colligitur. Subiungit idem. Coloris in labis pulchlitudo sanitamtem et bonam dispositionem signant stomachi. Haec ille.

Part VIII, 3: [...] Constantine maintains that people whose noses are straight have a tendency for heart failure. The breath which should pass through the nostrils is impeded by the straightness of the nose, which guides the breath through the chest. This often causes hypoxia and headaches.

Differetia VIII, 3: [...] *Concludit Constantinus, qui nares habet strictas, inclinat ad infirmitates pectoris. Supfluitas enim quae fluere debet per nares, per strictitudinem narium impeditur et fluit versus pectus. Unde et his saepius opitulationes narium incurrunt et capitis dolores patiuntur. Haec ille.*

There are also many passages in Glogoviensis' work that reference the theory of humours and temperaments. Like physiognomy, this school of thought, which was immensely popular in the Middle Ages, has its roots in the Antiquity.

Part V, 6: A small amount of hair [on the scalp] signifies dryness. A large amount of hair signifies favourable qualities of the brain. As Constantine maintains, the reason why people who were ailing lose hair during their recovery is that their brain was also ailing during their illness. It may be compared to the leaves in autumn. In autumn, the moisture travels down from the branches to penetrate the roots, and the leaves fall off. Similarly, when the brain absorbs all the moisture, the hair falls out. The aridity of the brain causes balding [...].

Differentia V, 6: Crinum paucitas in capite cerebri signat siccitatem. Crinum autem multitudo cerebri signat temperamentum et bonam complexionem. Unde et propter hoc dicit Constaninus quod homines, qui in convalescentia sunt, post infirmitatem capillorum habent casum propter exiccationem cerberi factam in infimitate. Adducit simile de foliis arborum in autumno. Cuius enim in autumno propter frigus circumstans humiditas et unctuosum quid est in ramis pellitur ad radicem, folia ab arboribus decidunt, sic et consumpto humido in cerbero crines defluunt. Unde et propter hoc id est siccitatem cerebri calvities generant in hominibus [...].

Part V, 8: People who go grey in their youth have brain damage, as there is a vast *discrasia* (mixing of fluids) in their brain, causing it to become frail. Such people, Constantine maintains, are garrulous, fickle and are vain of heart. Youthful greying may also have a different cause: it can be the result of experiencing supreme dread or anxiety, or conversely, of living in too lax a fashion. Mundinus concludes that hair growth is made possible by the substances circulating in our veins with the blood that nourish the hair. [As was mentioned before] the exact location of the humours in the body correlates with one's hair colour.

> *Differentia V, 8(ultima): Cui in iuventute advenit canicies celberi habet defectum,magnamque in cerrbero suo patitur discrasiam timeat talis lapum et infimitates graves quod ex cerebri debilitate eveniunt. Dicit Constantinus quod tales a natura signati existunt, suntque loquaces, inconstantes animi et vani cordis. Potest tamen etiam causa accidentali in iuventute homini advenire canicies ut propter timorem magnum, excura multa, et ex lascivia vitae. Concludit Mudianus quod capillorum crescentia est supfluum ex digestione tercia procedes quod fit in venis magnis unde et propter hoc homini crescunt crines quod diu nutritur et vivit, humorumque dispositio in homine ex crinum colore ut dictum est* [...].

Part VII, 12: Aristotle says that delicate lips suggest a sound intellect and are found in good-natured people. A man with pouting lips is offensive to others. As Constantine maintains, people who do not keep their lips firmly closed have strong gastric humours and are predisposed to aridity; they cannot keep warm [...].

> *Differentia VII, ultima (12): Inquit Aristoteles quibuscumque labia subtilia sunt et tenuitatem magnam habentia, sunt homines multarum cogitationum et boni ingenii et magnanimitatem tendentes. Quibus autem labrum superius cum gingivis multum extra eminent, homines sunt contumelias facientes displicentias et alios confundentes. Inquit Constantinus homines quibus fissurae fiunt in labris, humores acutos in eourm habent stomacho, siccitatemque ad infirmitates calidas tales sunt dispositi* [...].

Part IV, 6: A man with a sallow face has propensity for melancholy. He may be possessed of an active imagination. This classification follows Hali Habenragel, who writes that people with sallow-skinned faces are subject to the influence of Saturn.

> *Differentia IV,6: Color in facie hominis luteus, hominem significat inviduum melancholicum* [...] *fortitudinem fantasiae et imaginationis habet. Hali Habenragel in libro suo completo scribit* [...] *hominem enim lutei coloris de natura Saturni ostandit* [...].

Part IV, 7: In his *Phisionomia,* Aristotle writes of people whose preferred colour is white. They are merciful, kind and well-mannered [...]. On the other hand, people who take to the colour black are callous and their humours are more piercing. Constantine believes that the souls of the phlegmatics are more delicate [...].

> *Differentia IV, ultima (7): Inquit Aristoteles in Phisionomia sua quibus albus color est et purus, misericordes sunt et boni animi* [...] *Homienes autem qui ad colorem nigrum declinant plus rigidi sunt, humores habent acutiores. Inquit etiam Constantinus, homines flegmatici mollioris sunt animi* [...].

These passages prove that the face is of paramount importance in John of Głogów's treatise, and its appearance reflects character traits. There is a distinct lack of a new tendency that shifts the emphasis from the features of the face to facial expressions and emotions; nevertheless, signs of new schools of thought are demonstrably present in the work. As has been argued by the authors of *The History of the Face*, the sixteenth century saw the beginnings of an early modern project to improve upon human nature, which intended to shift the focus from blood ties to rhetorical ability. The foundation of this project was *civilitas* [civility], which entailed not only good manners and curbing the body, but also mastering words, as we can read in the works of Erasmus of Rotterdam. In the early modern concept of *civilitas*, which had its roots in ancient rhetoric, *eloquentia* [eloquence] and its sister discipline, *conversatio* [the art of conversation], were the factors shaping 'civil society'; both were considered expressions of humanity. John of Głogów's *Phisionomia* shows some early signs of this concept being gradually implemented, as in his work, the various physiognomic features are often matched to traits like eloquence and wisdom (*eloquentia et sapientia*). On many occasions, physical traits are also testament to a person's manners and their capacity to live within the constraints of society.

Half-way through the seventeenth century physiognomy was headed for collapse. This situation changed in the eighteenth century when physiognomy was revived by Johann-Caspar Lavater and Franz Joseph Gall. Their research, based on Ancient, Medieval and Renaissance treatises, restored the importance of physiognomy, through which new theories such as pathognomy and phrenology appeared and developed. In terms of phrenology, John of Głogów's *Phisionomia* was an outstanding phenomenon: in the third part of his thesis, the author writes about the relationship between the size of human head and the soul. Research for this piece (even if the description adheres to Bernard de Gordon – a French professor and doctor of medicine from the thirteenth century), made John of Głogów a precursor of phrenology three centuries before Franz Joseph Gall [Bentkowski 1814, p. 11].

Bibliography

1. Bentkowski, F., 1814. *Historia literatury Polskiey,* Warsaw-Wilno.
2. Birkenmajer, A., 1937. *Astronomowie i astrologowie śląscy w wiekach średnich*, Cieszyn.
3. Birkenmajer, A., 1956. 'Uniwersytet krakowski jako międzynarodowy ośrodek studiów astronomicznych na przełomie XV i XVI stulecia', in: *Odrodzenie w Polsce: materiały sesji naukowej PAN 25–30 października 1953 roku. Historia Nauki* Vol. II, 2, ed. B. Suchodolski, Warsaw, pp. 363–373.

4. Chojecka, E., 1969. 'Theorie und Praxis des Portäts der Früenaissance Die Phisionomia des Johann von Glogau (1518)' in *Wissenschatliche Zeitschrift er Fridrisch-Schiller-Universität Jena Geselschafts-und spracgwissenschaftliche Rheihe*, Heft 1, Jahrgand 18, 1969, pp. 177–180.
5. Coutrine, J. J., Haroche, C., 2007. *Historia twarzy. Wyrażanie i ukrywanie emocji od XVI do początku XIX wieku,* trans. T. Soboda, Wydawnictwo Słowo/ Obraz/Terytoria, Gdańsk.
6. Daston, L. J., Park, K., 1998. *Wonders and the Order of Nature: 1150–1750,* Zone Books, New York.
7. Davidson, A., 1991. 'The Horror of Monsters', in: *The Boundaries of Humanity. Humans, Animals and Machines*, eds. J. J. Sheehan and M. Sosna, University of California Press, Berkeley.
8. Dębicki, J., 2008. 'Wit Stwosz i Jan z Głogowa. O wpływie krakowskiej filozofii na sztukę Wita Stwosza' in: *Alma Mater* No. 109 (2008), Krakow.
9. Domański, J., 1978. *Filozofia i myśl społeczna XIII–XV wieku*, Państwowe Wydawnictwo Naukowe, Warsaw.
10. Ghersetti, A., Swain, S., 2007. 'Polemon's Physiognomy in the Arabic tradition', in: *Seeing the Face, Seeing the Soul. Polemon's 'Physiognomy' from Antiquity to Medieval Islam*, ed. S. Swaim, Oxford University Press, New York, pp. 309–329.
11. Jones, C., 2012. 'Presidential Address: French Crossings: III. The Smile of the Tiger', in: *Transactions of the Royal Historical Society,* Sixth Series, Vol. XXII, New York, pp. 3–36
12. Kuksewicz, Z., 1963. 'Główne źródła Komentarza do De anima Jana z Głogowa', in: *Studia mediewistyczne*, 4, Warsaw.
13. Porter, M., 2005. *Windows of the Soul: Physiognomy in European Culture 1470–1780,* Oxford University Press, Oxford 2005.
14. Sosnowski, L., 2010. *Sztuka. Retoryka. Fizjognomika. Studium z filozofii ekspresji*, Collegium Columbinum, Krakow.
15. Zwiercan, M., 1962–1964. 'Jan z Głogowa', in: *Polski słownik biograficzny*, Vol. 10/3, Wrocław - Warsaw - Krakow. pp. 450–452.

Latin texts

1. *Phisionomia hicinde ex illustris scriptoribus per venerabilem virum Magistrum Joannem Glogoviensem diligentissime recollecta,* Cracoviae 1518.
2. *Historia critica regnum Hungariae* (1793:258) – *Historia critica regnum Hungariae,* X (XVII), Budae 1793.

Katarzyna Trzeciak

White sculpted faces: Between the Medusa's gaze and prospopopeic blindness in Irving Feldman's *All of Us Here*

This article examines the sculpted faces in poems by American author Irving Feldman. Starting from Michel Serres' philosophical investigations into sculptures as representations of the dying human body, the article goes further into the myth of Medusa and the question of the sculptural gaze. Feldman approaches George Segal's sculptures from a literary perspective to raise the dialectical issue of solitude, the opacity of matter and the viewer's gaze. Viewers in the poems of *All of Us Here* have to cope with the sculptural opacity of matter, and to rethink the status of their human gaze.

Keywords: sculpture, statues, Irving Feldman, Medusa's gaze, monument

What is a statue? Is it a piece of material, like wood, bronze or stone? A human double? The word 'statue' holds a broad range of meanings. In the philosophical anthropology of sculpture elaborated in Michel Serres's *Statues* we read that statues are a supplementary population representing the people in human history [Serres 2015, p. 5]. Actually, the statue represents, and roots the dying, entropic human body, revealing the presence of a corpse, of our decaying materiality. It doubles the cadaver, not the living body. 'This means that the statue conceals our fear of the living body as well, our anxiety over its wastes and sensitivities.' [Gross 2006, p. 21].

The reality of the Gorgon's face

The sculptural figure that embodies such qualities is the Medusa, or the Gorgon (never mentioned by Serres), a creature whose gaze turns the living to stone. The most peculiar fact about the Gorgon is, however, her visual representation: the frontal view of her face.[1] This is the face the ancient Greeks gave to everything alien, other, chaotic and destructive, especially those cruelties and disharmonies

1 Jean-Pierre Vernant recalls Medusa's appearances from the seventh century onward in all the various forms to be found in Corinthian, Attic, and Laconian imagery. He also notes that the only Olympian god to be represented facing front is Dionysus. See: Jean-Pierre Vernant, Pierre Vidal-Naquet, *Myth and Tragedy in Ancient Greece*, trans. J. Lloyd (New York: Zone Books, 1990), p. 191.

that disrupted what Greek culture tried to define as the domain of the human. Such disruptions could include Dionysian ecstasy as well as the frenetic rage of battle, the music of the flute as well as the noise of clashing bronze [Gross 2006, p. 24]. The face of the Gorgon was also the face of the dead, the face of the demanding revenant or ghost.

But probably the most important and culturally remembered characteristic of Medusa's gaze is the soft-sculpted mirror. Jean-Pierre Vernant suggests that the Gorgon mask reflects our own terrified gaze at what we cannot bear to apprehend. 'It is your gaze that is captured in the mask', Vernant insists; 'what the mask of Gorgo lets you see, when you are bewitched by it, is yourself, yourself in the world beyond, the head clothed in night, the masked face of the invisible that, in the eye of Gorgo, is revealed as the truth about your own face' [Vernant 1991, pp. 137–138]. The Gorgon's gaze connotes the dangerous and horrific aspect of reality.

This analysis begs for juxtaposition with a Lacanian perspective. When we combine it with the Lacanian concept of the Real, we see that the horror of the Medusa is a return to the formless ground of reality. Medusa's gaze exposed, unmediated, is a gaze that can swiftly turn us into stone and deprive us of spirituality. Under this gaze human existence is reduced to raw material. But does this mean mere death? Or maybe materiality constitutes a different kind of life? And if so, how can we interpret the material, sculptural life of the statues and of the human subject confronting them?

The Image as a cadaver

Irving Feldman, an American poet and the author of *All of Us Here,* raises some questions about the gaze between human and sculpture in his collection of poems created as a series of meditations on the work of the contemporary sculptor George Segal. Segal's sculptures generally consist of one or more full-size human figures, made of stark white plaster, usually set not on pedestals, but on the bare floor, spectrally inhabiting what at least appear to be authentic fragments of a world of things and signs. 'One can think of tableau sculpture – such as the work of George Segal [...] as theatrical, although no internal mechanization impels the sculptured actors to "perform" in time. It is rather the viewer's movement as he walks around the sculptural diorama, or takes time to interpret the narrative meaning of the various details of the tableau, that endows these works with dramatic time' [Krauss 1981, p. 221–226]. Interpreting the narrative meaning of Segal's statues is the essence of Feldman's poetic project. Each poem attempts to make sense of the amalgam of opacity, solitude and temporal dislocation of these figures, as well as their peculiar connection with the viewer's gaze.

> It's true they loom like revenants in their scenes,
> Old blanks, cut-outs, massive absences
> who want to climb back into the world which keeps on going,
> misfits trying to get at least something right,
> who mime with dogged, solemn single-mindedness
> this one's lopsided frown, that one's wrinkled suit.
> Is this why suddenly we feel it disloyal
> of us if we smile, stretch, straighten our clothes?
> -But it is only now we notice it.
> They have no eyes or their eyes are closed.
> [Feldman 1986, p. 6]

The last line suggests that Segal's sculpture embodies death. This reading of statues thus enacts Maurice Blanchot's suggestion that all human images 'have no guarantee but a cadaver' [Blanchot 1981, p. 83]. The subject of Feldman's poems sees white figures as dead people: 'Sadly, it's the dead themselves they resemble' [Feldman 1986, p. 8]. The juxtaposition of death and image does not appear to be incidental. In the philosophical context of Blanchot's reflections, the cadaver is an image that 'no longer entertains any relation with this world, where it still appears, except that of an image, an obscure possibility, a shadow ever present behind the living form which now, far from separating itself from this form, transforms it entirely into shadow' [Blanchot 1982, p. 257]. According to Blanchot, the most frightening aspect of the image-cadaver dialectic is the fact that the cadaver is an image which should have no connection to the world of the living, though it does. The cadaver endures in an image; every image is already a cadaver and transforms every living form into a shadowy form of life-in-death.

Feldman writes about sculptural 'revenants' who inhabit the strange space between life and death. They exist beyond the human world, but because of their awkward, white materiality, they still haunt reality as liminal beings. The concept of the 'revenant' can be hardly related to materiality. It is something intangible, the opposite of a material being. But in Feldman's poetic meditation it is precisely the materiality of statues that makes them spectral in human reality.

The white materiality of faces

Segal's sculptures are made of white plaster, with white faces and hollow eyes. Whiteness does not mean a transparency that allows the viewer to see the inner life of the material. A white face looks more like a mask which hides the truth of the emotions and the spiritual depths. There is no difference between the face and the mask, so the white surfaces of Segal's faces become pure materiality that cannot be easily read as a sign of hidden substance. Segal goes beyond an understanding of the face as a surface or a screen for non-material, spiritual values. His sculptures disrupt clear communication between these two fundamental levels of human existence and show that there is only a white, opaque surface that we can try to read. According to Rosalind Krauss, this communication between the surface and the depths was eliminated by Rodin, who left us 'with the gestures that are unsupported by appeals to their own anatomical backgrounds, that cannot address themselves logically to a recognizable, prior experience within ourselves' [Krauss 1981, p. 27]. Segal tries to sustain this rupture, and Feldman shows how this strategy affects the viewer's gaze.

The whiteness of statues blinds and paralyses the beholder. White surfaces emanate the blind and mute horror of pure material – all that is left of our heroic and eschatological dreams. 'They show us our frozen, somewhat banalized terror, our 'petrification' in the face of those dreams, as well as in the face of collapse' [Gross 2006, p. 27].[2] The materiality of the statues is the 'white lie' of our culture; they embody the fact that the face is not a transparent surface through which we can see the truth of humanity. There is no soul beneath, though they do indeed have some life:

> A hidden life-turned low to endure-
> persists in there, and contends with us,
> with itself, with everything
> for the meaning of the life.
> And will not let us rest . . .
> [Feldman 1986, p. 33–34]

'The hidden life' has one main goal: to endure. Segal's statues more resemble zombies than revenants. The white sculptures are the remains of human eschatological fantasies. They are bodies without subjectivity, representing the no-man's land between the social, symbolic, meaningful life, and bare, biological existence. It is the space of the 'neighbour with whom no empathic relationship is possible' [Mullen 2014, p. 49]. From Slavoj Žižek's point of view, the zombie, or the undead, is a fantasy about the subject's condition in relation to the symbolic network of reference and meaning. Subjectivity finds its meaning, identity, and purpose be-

---

2 Gross claims that 'Feldman troubled our dreams of the commonplace. Dreams of a secular, human and communal life by the fact that our "fellow citizens" now include the kind of life-size cartoon figures that walk through Disneyland, the living corpses of Auschwitz, and the ashen silhouettes of Nagasaki' – Gross 2006, p. 28.

cause it is entangled in the symbolic order. The biological, material dimensions of life have to be subjugated by cultural roles, metaphors, and narratives. And the most crucial question, according to Žižek, is what happens when we no longer recognise ourselves within the symbolic network of meanings. Is it possible for us to become bodies without subjectivity – mere life without a language in which this life can be understood or communicated? [Mullen 2014, p. 49].

Feldman's poems stage exactly this moment of confrontation with material bodies without a symbolic network of references, when the white, frozen remnants shorn of our narratives come to life. The statues have an 'extra life, like *more* life in the veins,' [Feldman 1986, p. 25] but 'What welcomes us here is only ourselves in a sudden vision of ourselves' [Feldman 1986, p. 7]. This is probably the crucial insight in Feldman's collection. The white plaster faces have only a Medusa-like gaze that reflects ourselves gazing at what we cannot bear to look.

White materiality is an impassable barrier for the beholder, and on the surface of sculpted faces we see only our own terrified gaze. With no cultural narrative to explain Segal's sculpture we are confronted with the reality of objects that stare directly at their viewers. In Hal Foster's interpretation of the Lacanian 'Real', this concept is connected with the object-gaze. Because human subjects have access to a symbolic order, they can protect themselves from the object-gaze. The symbolic order draws from a cultural reservoir of narratives, artistic conventions and schemata of representation. Together, these build a screen that mediates the object-gaze for the subject. 'In this way the screen allows the subject, at the point of the picture, to behold the object at the point of light. Otherwise it would be impossible, for to see without this screen would be to be blinded by the gaze or touched by the real' [Foster 1996, p. 140].

The viewers in *All of Us Here* try to cope with being blinded by the sculptural gaze. They project different symbolic meanings upon plaster statues in order to prevent themselves from being turned into bare, meaningless materiality. The experience of life without subjectivity, of life devoid of any cultural ability to communicate, has to be transformed into language, otherwise it is impossible to understand:

> They're like completed destinies that are at the same time
> Just poor people who couldn't get out of the way,
> Whose names have been exalted into allegory:
> Exile, Homeless, Refugee, Unknown, Mourner, Corpse.
> [Feldman 1986, p. 32]

Viewers defend their integrity from unmediated contact with the statues' bare materiality by endowing the figures with allegorical meaning, which can work as protection against the decay of subjectivity. Allegorical meanings draw from human symbolic desires to give redemptive moments of transcendence to existence, and to paint the white faces with traces of deeper significance, to cover up the terrifying fact that bare material life has nothing to defend.

Vernant clarifies why Segal's sculptures have a petrifying effect. 'The Gorgon marks out the boundary of the world of the dead. To pass that frontier is to become, oneself, beneath her gaze and in her image, what the dead are – empty heads without strength, heads shrouded in night' [Vernant 1990, p. 192]. The Gorgon is the horror of 'blurring all categories' used by human subjects to structure reality; she embodies a radical rupture in the symbolic order, and for this reason, she must be turned into a metaphor or an allegory.

If Segal's statues have a Medusa's gaze, the beholders can see only their own fear in the statues' eyes. The white faces must be included in the higher order of significance because this is the only guarantee of a meaningful human death.

> They are dying to be the letter itself:
> immaculate, and perfect in form, minute,
> not ever again to be read into,
> and beyond whiteness white, sole, invisible.
> [Feldman 1986, p. 9]

Feldman presents the moment of dying as a petrification in language, in conventions of a symbolic order. Statues become illegible letters that have no meaning besides their material perfection. And this death requires eschatological explanation, so the beholders try to invent narratives to give meaning to the lifeless void.

The line 'We will arise from these tombs of hoarfrost and ash' [Feldman 1986, p. 9] indicates the possibility of breaking the dominance of the symbolic. This possibility lies in looking at a white face and trying to perceive our own condition as material beings. It is an image of an ethical act that rejects the reigning order of reference, meaning and value. This appearance of something that resists the surrounding world without reliance upon an immutable beyond and in the fullest embrace of its own contingency and truth is the 'pure surface' which is 'eternal' only in the sense that it wholly resists the dominant trend of the time (or historical horizon) in which it finds itself.

White writing

Feldman's literary attempt to present Segal's statues emphasised their surfaces in different ways. One of them is the peculiar transposition of the whiteness of plaster into writing. David Huerta notes that the whiteness of both Segal's figures and Feldman's writing does not signify the void or a vacant place in the world, it is only the beginning of a movement, of unfolding, which develops the poetic discourse of *All of Us Here* [Huerta 1992, p. 42]. Whiteness or colourlessness appears in Roland Barthes's writing in the context of *writing degree zero.* It is precisely the whiteness of language that is the starting point of modern literary texts, as well as their conclusion. Barthes' definition of writing intended to show what writing is not. It is not language, that ordered, collective, apparently 'natural' (but actually historical) horizon of social experience; nor is it style, the inevitably crude residue of an author's biology (body) and biography (past) [Evans 2000, pp. 9–12]. Writing emerges as a third term between these two forms, as a possibility (not a destiny), as a value, not a fact. A mode of writing is an act of historical solidarity (Evans) with a counter-factual community whose value is 'freely produced language'.

Feldman's poems constitute a specific community built around a kind of non-identical identity:

> The mirror they hold up to us
> is tempered with a gentle curve of revery.
> And whether it's they or ourselves we see
> doesn't much matter. Whoever's there-presence
> and person and apparition and absence,
> the once, the actual, the not yet, the never-
> we don't want to be halted by identities,
> we want to go on becoming in wonder.
> About these figures we don't ask, 'Who are they?'
> We ask, 'Who, who is it they remind us of?'
> [Feldman 1986, p. 36]

Community beyond universal meanings and identities is another strategy for making a different kind of contact with the statues. 'Who are they?' is a historian's question, part of the domain of knowledge that ascribes abstract names (like allegories) and generates a sense of alienation from concrete persons and their experiences. So instead of believing in knowledge, Feldman proposes we ask about the real meaning of the statues, which he claims lies in standing together with them in a fictional scenario: 'not quite sure if we recognize anyone, even who is still alive. We and the statues, caught in shared revery, may hope only that some of the faces on the squares of treated paper will "augment in our considering"' [Gross 2006, p. 30].

Feldman believed that between the bare materiality of Segal's statues, their Medusan gaze, and allegorical mode of defence against petrification, one can find a memory-based form of personal communication that is never full and complete.

> But even there in the land of the dead,
> lost not in time but out of it, in timelessness,
> among the quiet, whited and nonchalant crowds,
> where nothing moved but our shadows' vague passage,
> a steady gaze of recognition from afar,
> the happy affection glowing in those eyes
> as we approached, announced who lit our way
> and who it was there among many:
> it was they, themselves, and we were ourselves!
> [Feldman 1986, p. 38]

The most crucial task for the beholders is to understand that the petrified faces of Segal's statues should not be read as universal symbols or allegories, nor as terrified zombies with material form. In the poem 'Did they love one another?' Feldman shows that only in the act of constantly inventing fictional scenarios can we, the beholders, draw nearer to personal stories of the statues that can come alive in our idiomatic fantasies. It is not about assigning allegorical significance, creating metaphors and associating each individual sculpture with a hermeneutical interpretation. It is rather that the beholders have to try to open themselves to an unstable community of beings.

The last poem of the cycle, 'They say to us', catches human-beholders and statues in a moment of looking at old photographs. The statues explain the faces in the photographs and the people pass them from hand to hand. Even if the people fail to recognise the identity of the faces, they have to listen to the sculptures' voices and try to imagine their own connection with the unfamiliar images. According to Feldman, only this attempt at communication between sculptures (as an embodiment of the past which is never fully accessible, but has to be reinvented) and the people observing them is ethically fair, insofar as it is the only way to avoid both allegorisation and materialisation (restricting the statues to their materiality).

However, the title of the last poem, 'They say to us', suggests an important concept that lets us deconstruct the naïve faith in the idiomatic, non-identical identity asserted by Feldman. The poet believes that, in an imaginary scenario, we, the readers/beholders, can give language to the abstract statue, and through language, restore their private histories. Therefore, Feldman's poetic concept and final conclusion involves lyrical prosopopeia: 'the movement of bestowing a face or voice on an absent or dead speaker in order to stimulate his momentary narrative presence' [Richter 1998, p. 424].

Falling in prosopopeic love

'The figure of prosopopeia', Paul de Man argues, is 'the fiction of an apostrophe to an absent, deceased or voiceless entity, which posits the possibility of the latter's reply and confers upon it the power of speech. Voice assumes mouth, eye and finally face, a chain that is manifest in the etymology of the trope's name, *prosopon poien*, to confer a mask or a face (*prosopon*)' [de Man 1979, p. 926]. Prosopopeia is the act of bestowing a face in a linguistic sign, on the one hand, and on the other 'by making the death speak, the symmetrical structure of the trope implies, by the same token, that the living are struck dumb, frozen in their own death' [de Man 1979, p. 928]. In this transformation we see the ambition of language-users to create something fixed or whole; at the same time, it reveals the failure and violence of this ambition. From the outset, the statues that languages create appear to be broken, mutilated, and defaced [Gross 2006, p. 151]. When de Man writes on Rousseau's *Pygmalion*, he notes that the sculptor experiments with different strategies of facing the 'monstrous goddess' embodied in the immobile statue. The fictitious artist tries allegorical reading, identification, aggression and adoration, yet none of these strategies succeed. All of them only reveal an endless chain of substitutable figurations.

According to de Man, then, there is no escape from the contaminated antinomies of self and other, death and life, art and nature. 'For de Man, all attempts to use language to give form or meaning to the world - to describe a cause, locate an origin, posit a subject, reconstruct a lost wholeness, to mirror a reality - turn our words, and we who use them, into statues, into monuments and tombs' [Gross 2006, p. 151].

In all the poems in *All of Us Here*, Irving Feldman uses similar strategies which, he believes, might lead the reader/beholder to understand the meaning of statues. The final use of prosopopeia, however, is intended as an act of communing with the statues, beyond both allegorisation and materialisation, revealing no more than another level of substitution. The imaginary faces and voices given by language are not unmediated, authentic presences, but a simulacrum of the face made by a discourse calling upon us for authentication. Feldman believes that such authentication is possible in the act of creating individual, private identities for statues. He does not see prosopopeia as a phantasmagoria of immediacy and presence, and ignores the fact that by making the dead speak, the living become mutilated and frozen in their own death.

In the final lines, 'All of us here / are deeply satisfied' [Feldman 1986, p. 41], it seems that, after failures at communicating with the statues, there is evident success in the contact between humans and statues. Feldman is sure that language is not an instrument of alienation, but one that is transparent, that makes unmediated, authentic communication possible. An idiomatic, private language is the opposite of the allegorical, depersonalising mode of reading, Feldman seems to claim. Yet this language of private experience does not reveal 'true faces', because at the very root of those experiences lies prosopopeia itself. Feldman uses prosopopeia, but constantly masks its mortal dimension of 'de-facement'; he fears to confront it.

'To face the prosopopeia, then,' writes Richter, 'is to allow for the experience of an aporetic structure, that is to say, the simultaneity of recognizing in this structure's simulacrum of the face the contours of a discourse calling upon us for authentication or verification - our counter-signature - and of resisting the temptation to succumb to its ultimately unsustainable phantasmagoria of immediacy and presence.' In Feldman's poetic discourse this aporetic structure is hidden and covered by a faith in the immediacy of language. Prosopopeia allows the poet to give a face to the statues, but it must be hidden as an act of simulacrisation.

Betraying ornament

By using prosopopeia, Feldman also tries to redefine the meaning of a monument. Segal's sculptures, as figures from some indefinite past, cannot be connected with any particular moment in history (if this were possible, we would face an allegory of the historical past, which Feldman wants to avoid, as depersonalisation). The viewer cannot just inquire into their origin or identity because they do not pertain to any stable meaning in the past. Instead, we should give them language and let them tell their private story, which is not fully understandable because it has no stable roots that are universally recognised in the symbolic language of events and times. 'Monument's truth', Gianni Vattimo proposes, 'in both senses of the genitive: the authentic meaning of monument and truth-as-monument – only grants itself when monument unequivocally ceases to be associated with foundation. As long as foundational thought holds out, as long as metaphysics lasts, monument is merely a metaphor for foundation, an allegory of it, and vanishes as soon as thought relates to the proper meaning; it is then, perhaps, only a rhetorical ornament, a way of conveying an everlasting truth concerning existence and values, which it provides with an "aesthetic" coating. Within that horizon, only those monuments really last that survive such translation into their "proper" meaning, such demythologizing, one might say' [Vattimo 1995, p. 43].

Feldman seeks a way to avoid a foundational perspective in his interpretation of Segal's statues. He wants to go beyond metaphysical thinking, which foists allegories onto the statues. As he demonstrates, allegories dissolve faces by giving them totalising meanings. Overwhelmed by an allegory, individuality vanishes, leaving only empty words like 'Exile', 'Refugee', or 'Mourner'. That is why Feldman uses prosopopeia to let the individuality speak for itself. Yet he also fails to escape a desire for their 'proper meaning', although he locates it on a different level of communication. Vattimo writes about the dissolution of metaphysical, foundational thought that is, at the same time, an act of freeing an ornament. Feldman, who stays blind to the aporetic structure of prosopopeia, persuades us that it is possible to think and to express foundational experiences in the language of private conversation with statues. The ornamental character of a monument seems insufficient, somehow incompatible with the truth of Segal's sculptures. Therefore, we, the beholders, have to find authentic faces in the empty eyes of the statues. Trying to demythologise the symbolic language of the monument, Feldman falls into another myth - the transparency and authenticity of a private language. And to give the readers/beholders satisfaction in meeting the statues, he conceals the fact that behind authenticity of communication stands the negative power of the linguistic sign.

At the opening of the present essay I mentioned the gaze of the Medusa in Feldman's collections of poems. To protect the beholder's integrity, the subjects from *All of Us Here* tried different methods of pacifying the Real. Prosopopeia, an act of lending voice and face to sculptural figures, is the final stage of making the horror of white, mutilated statues more familiar to the living. However, in Feldman's poems, the living stay blind to the fact that they are already in the same position as the dead sculptures. And that their living faces are only monuments to their own oblivion. Any attempt to undo this process merely repeats it. This is perhaps the single most important image inscribed on the white materiality of Segal's sculpted faces.

Bibliography

1. Blanchot, M., 1981. *The Gaze of Orpheus and Other Literary Essays,* trans. L. Davis, ed. P. Adams Sitney, Station Hill Press, New York.
2. Blanchot, M., 1982. *The Space of Literature*, trans. Ann Smock, University of Nebraska Press, London.
3. Evans, S., 2000. *After a Fashion: Reading Roland Barthes Today, Poetry Project Newsletter* February/March 2000: pp. 9–12, accessed at: http://thirdfactory.net/barthes.html
4. Feldman, I., 1986. *All of Us Here,* Penguin Books, New York.
5. Foster, H., 1996. *The Return of the Real,* The MIT Press, Cambridge.
6. Gross, K., 2006. *The Dream of the Moving Statue,* The Pennsylvania State University Press, Pennsylvania.

7. Huerta, D., 1992. 'Unfolding the Figure: On the Poetic Space of Irving Feldman', in *The Poetry of Irving Feldman: Nine Essays*, edited by Harold Schweizer (Lewisburg: Bucknell University Press, London and Toronto: Associated University Press)
8. Krauss, R., 1981. *Passages in Modern Sculpture*, The MIT Press, Cambridge.
9. Man P. de, 1979. 'Autobiography as De-facement', *MLN*, Vol. 94, No. 5, Comparative Literature.
10. Mullen, G., 2014. *Adorno, Žižek and the Zombie: Representing Mortality in an Age of Mass Killing, Journal for Cultural and Religious Studies* Vol. 13 No 2.
11. Richter, G., 1998. 'Face-Off', *Monatshefte*, Vol. 90, No. 4.
12. Serres, M., 2015. *Statues: The Second Book of Foundations,* trans. R. Burks, Bloomsbury, London.
13. Vattimo, G., 1995. 'Postmodernity and New Monumentality', *RES: Anthropology and Aesthetics*, No. 28.
14. Vernant, J.-P., 1991. *Mortals and Immortals*, ed. F. I. Zeitlin, Princeton University Press, Princeton.
15. Vernant, J.-P., *Myth and Tragedy in Ancient Greece,* Zone Books, New York.

Adrianna Alksnin

Me, myself and my face: Pirandello meets Gombrowicz

This article is an attempt to compare an Italian and a Polish modernist novel: *One, No One and One Hundred Thousand* by Luigi Pirandello (1926) and *Ferdydurke* by Witold Gombrowicz (1938). The novels present the modern subjects' struggle to regain solid identities, to resist the fluidity of the modern world and the oppression of social rules and regulations. Both writers emphasise the performative dimension of subjectivity, which should be perceived as inter-subjective, as it is born and developed in relation to another subject, and perpetually vulnerable to its influence.

Keywords: self, identity, subject, face, modernism

The following article is an attempt to compare two novels, of Italian and Polish modernism: *One, No One and One Hundred Thousand* by Luigi Pirandello (1926) and *Ferdydurke* by Witold Gombrowicz (1938). Both texts can be read as documents of disintegrating Identity, owing to social structures that set the confines of participation. Although the two writers' diagnoses take different directions and search for different solutions, they both point at a factor that plays a crucial part in forging an identity in society: the face. Both novels investigate the role that the face plays in constructing the Self – the face and its expressions as something deeply personal which cannot be seen by its owner, and the face perceived by others, who build their relations with a person on the basis of their face. They also (metaphorically) examine the sense of self through the face – how physical changes can affect the personality and how personality disorders can change a face (the internal invading the external).

In search of the true face of Vitangelo Moscarda

Both the main character and the narrator telling the story, Vitangelo Moscarda is a citizen of the small rural town of Richeri. Moscarda, the thirty-something son of a banker, is a typical 'Man without qualities': despite his intellectual and financial abilities, he cares little for his career, or for practical or cultural affairs. He is rather lazy and lives off his father's fortune: the bank and numerous properties. He is not interested in the financial mechanisms of the bank he owns, leaving all decisions to his plenipotentiaries: Mr. Quantorzo and Mr. Firbo. In other words, he lives an idle life.

One day, his situation changes entirely: while he is examining his face in the mirror, his wife Dida remarks that his nose is 'drooping' on the right. Rather surprised, he admits he had never noticed this defect. Worse still, his friends, when asked, confirm they have noticed his nose drooping. This apparently meaningless remark triggers a stream of questions: 'Who am I to Others?', 'Who am I to Myself?', and therefore, 'Who am I?'. This last question may be paraphrased as 'What does my REAL face look like?' as in the Pirandellian novel it is the face, as the manifestation of the Self, that is the external avatar of internal qualities. What terrifies Moscarda is the realization that he has no true face of Vitangelo Moscarda, but that everyone has his or her own face of Vitangelo, which is different from what he thinks he is.

This discovery leads Vitangelo to a state verging on madness. This state, which he calls a 'malady', can be understood in two ways: firstly, as a neurotic obsession that he develops as a reaction to his discovery; and secondly, as a disability that constitutes the personality as such. Supposing one can function as if one were a strong and 'healthy' subject, as long as one does not realise one is insane – in other words, insofar as one does not recognize one's inability to be self-determining, independent and not mediated by the gaze of others. In this sense, this neurosis is a second-tier malady which occurs when one realises this pre-existing meta-malady, this gap which constitutes all existence as such.

Such deliberations lead Moscarda to the following conclusions: if each and every citizen of Richeri, including his wife Dida, has his/her own image of Moscarda, this 'one' Face is multiplied by one hundred thousand, and the real Moscarda does not exist – he is 'no one'. From here on, Vitangelo tries to find his own true face, free from gazes that turn it into an 'untrue face'. In other words, he tries to discover 'who I am when no one is looking at me, not even myself'. The process of being liberated from the imposed Face is thwarted, for the figure his wife traps him in is a lazy and simple-minded loser, whom she calls 'Genge', and the citizens of Richeri perceive him as a cynical usurer. Moscarda could not object when he did not know who he really was. He was also in the process of putting a mask on his Face, or as Pirandello writes, 'fitting into a Form'.

'Form' in the Pirandellian novel is quite similar to the definition that emerges from *Ferdydurke*. It is the tool that enables the subject to perform an operation instantiating objects for perception and cognition; an operation that inscribes a *signifié* into an object, turning it into a meaningful sign – a *signifiant*. In these terms, the form is responsible for the enabling communication between two subjects, for understanding the world and another person. However, this relation is a cruel and violent one, based on a desire to dominate the partner in dialogue. In Gombrowicz's words – to impose a Form on the encountered subject and to drive him to act as one desires.

To escape an imposed Form, one needs to create his own Form, but as Pirandello suggests, there is always more than one Form rooted in others' perceptions. Unfortunately, integrating various Forms and Faces into one true Face is nearly impossible. The destruction of Forms and images someone else imposes on us may lead to a great misunderstanding, as the end of Pirandello's novel depicts. Moscarda, who tries to defy the expectations of Richeri's citizens, including his plenipotentiaries and his wife (he plans to close the bank, to finish a university degree and donate an estate to local sculptor Marco di Dio, whom he just recently evicted), leads them to a simple conclusion: Moscarda must have gone mad. Although Moscarda finally believes he is a 'real person', not just an avatar of other people's judgements and expectations, the threat of incapacitation is near. Help comes from a local bishop, who arrives at a solution: Moscarda should donate all his belongings to the poor and start a new life in a shelter in the suburbs of Richeri. Free from gazes and the influence of others, Moscarda's peace of mind is restored. Utter solitude is the key. He cannot see his own face, but his face cannot be seen by others. It is not clear whether Moscarda finds his own face, his identity, but one thing is certain – he does not care any more. If he is cured, it is not from the discontentment of being 'unwhole' and somehow 'defective', a multiplied subject; but he *is* surely cured from a neurotic longing for the integration of the Self. Just as being sane means accepting socially constructed rules and conditions for participating in the public sphere, agreeing to be an incomplete subject, always in relation to other subjects and their perceptions, stands for mental health, which Moscarda does not exactly regain, but rather decides to ignore.

Ferdydurke: from face to mug

The main character, Józio, a thirty-year-old writer and the narrator of *Ferdydurke*, suffers from a 'condition' not unlike Vitangelo Moscarda's. In the few first pages of the novel, Józio has just awoken, and is waging a war with his imaginary double, who seems to mock his gestures and therefore – his writings. This prompts Józio to deliver a speech on the mission of the writer and the necessity of being original and unique:

> I was left alone but actually not alone – how could I be alone when I wasn't even there, I had no sense of being there, and not a single thought, gesture, action, or word, in fact nothing seemed to be mine, but rather it was as if it had all been settled somewhere outside myself, decided for me – because in reality I was quite different! And this upset me terribly. Oh, to create my own form! To turn outward! To express myself! Let me conceive my own shape, let no one do it for me! [Gombrowicz 2000, p. 14]

Just as he tries to turn his 'manifesto' into writing, to find his own symbolic, textual signature – a meaningful representation of his deepest self, the first obstacle appears, in the person of Professor Pimko. This sudden appearance of a professor suggests two things; firstly, that Józio cannot create his own Form as long as he does not know who he really is as a subject, and secondly, that if he wants to create his own Form, he needs to fight for it. The process of 'Unmoulding' the Form, or maturing to become a subject, begins. But in this particular scene Józio is powerless – the professor has total control over him as he mumbles gibberish incantations: 'chirp, chirp, chirp…'. With his school authority, Pimko 'moulds' Józio into a pupil, 'green' and helpless, deals him a *pupa*. Although 'pupa' does not exist in English, the translator opted to use it instead of a translation, as it has metaphorical meaning. 'The Polish word *pupa* (pronounced "poopa") presented a special problem. It means the buttocks, behind, bum, tush, rump but not one of these (nor any others I considered) adequately conveys the sense in which Gombrowicz uses "pupa" in the text. While the "mug" is Gombrowicz's metaphor for the destructive elements in human relationships, the pupa is his metaphor for the gentle, insidious, but definite infantilising and humiliation that we inflict on one another. We made the decision to stay with the Polish word'. [Gombrowicz 2000, 'Translator's Note', p. xix]. The operation of dealing him a *pupa* can be read as a manifestation of symbolic violence against Józio.

Gombrowicz's relation to Form can be found in an article published soon after the release of *Ferdydurke*. He wrote the following in *Wiadomości Literackie*: 'The main issue in *Ferdydurke* is form. We live in an epoch of rapid change and swift development, when all the old solid forms are cracking under the pressure of life. […] The central emotion of my book derives from the necessity of finding a form for what remains immature, pliable, and underdeveloped in human nature, and a groan at the helplessness of this predicament' [Gombrowicz 1937].[1]

This erosion of solid and secure forms of social and cultural participation was caused by various factors, and was a major issue in European Modernism. We might borrow Richard Sheppard's three basic aspects of this issue thematised in early twentieth-century literature. These are: a changing sense of reality (a change in previous 'conclusions about the nature of physical reality' in the wake of new discoveries in quantum physics and the theory of relativity), a changing sense of

1 *Głównym problemem Ferdydurke jest problem formy. Żyjemy w epoce gwałtownych przemian, przyspieszonego rozwoju, w której dotychczas ustalone formy pękają pod naporem życia.* […] *Potrzeba znalezienia formy na to, co w człowieku jest jeszcze niedojrzałe, nieskrystalizowane i niedorozwinięte, oraz jęk z powodu beznadziejności tego postulatu oto naczelna emocja mojej książki.* [Gombrowicz 1937].

human nature (with the popularity of psychoanalysis and theories of the unconscious) and a changing view of the relationship between Humanity and Reality (the arbitrariness and relativism of language, the narrative nature of History and science, and a questioning of the objective paradigm of cognition, the expansion of subjectivity) [Sheppard 2000]. The emancipation of women, changes in the structure of society (the growth of the working class, migration from the rural areas to cities, the secularization and equalization of society), and the relaxing of public morality that swept through Europe at the turn of the century are also noteworthy. All these factors inspired the search for a new formula for subjectivity, free from the cracks in the Cartesian formula found by Freudians; a modern subjectivity that would face the challenge of modern world [Taylor 1992]. In speaking of Gombrowicz, we should also mention the historical context. In 1918, Poland regained political independence after 123 years. The question of a modern subjectivity was, unsurprisingly, also a question of national identity – finding a new 'Polish form' of being. This was the topic of numerous texts, manifestos, and deliberations in the interwar period. One should not be surprised that these attitudes were often contradictory, aligning themselves with the historical models of Polish Romanticism, or rejecting them, claiming the necessity of creating new (modern) ones, free from the old phantasms.

Though he adheres to the framework of the national, Polish aspect of subjectivity, Gombrowicz raises this discussion to a higher level [Chwin 2005]. It is existence that interests him, and how the existence of an individual overlaps with nationality. Form, as he understands it, is 'the source of symbolic oppression and a device to rescue us from the anarchy and chaos of "unformliness". It is form that enables us to participate in the first, primordial conflict between man and his form, just as painful as it was ages ago. It expresses human hatred, fear and shame against formlessness and anarchy. It is not a pamphlet, polemic nor a critique, it is the groan of an individual who staves off "disorder", who vigorously demands form and hierarchy, while perceiving that each form winnows and imprisons him' [Gombrowicz 1937].[2]

Form itself is not perceived as a source of trouble, but it may turn into one when it is imposed by force or when it is somehow inadequate, anachronistic or imposed

2 [*Ferdydurke* – przyp. A.A.] *Usiłuje wypowiedzieć i to jest jej punkt, najistotniejszy odwieczny konflikt pomiędzy człowiekiem a jego formą, konflikt równie bolesny dziś jak przed wiekami. Jest wyrazem nienawiści, lęku, wstydu człowieka wobec bezformia i anarchii. Nie jest, ściśle biorąc, pamfletem, polemiką, krytyką, ale po prostu jękiem jednostki, która broni się przed rozprzężeniem, gwałtownie domaga się hierarchii i formy, a zarazem widzi, że każda forma uszczupla ją i ogranicza* […].

by social convention; in other words, when Form is extraneous, not derived from the Self. Being caught in a Form against one's will is a source of discontent. An individual trapped in this situation seems to have two choices: either let the Form mould him, forming his face into a 'mug' and tolerating the imposition, or destroy the Form. According to Gombrowicz, laughter is the ultimate weapon against Form. When caricatured and turned into its own parody, Form loses its power to mould, but the freedom gained by the subject is ethereal and momentary. Soon, an individual is caught by another form. And then by another, and so on: 'Man, claimed Gombrowicz, exists only in relation to other men. When left alone he is a shapeless, changeable lump of contradictions. He receives Form – his self, his character, his multiple public images – in myriads encounters with others. [...] Thus created, Form is often ridiculous and awkward. More importantly, however, it is the only thing modern man can hold on to when describing his identity. There is no authentic self that can be liberated from Form. Man is always artificial, and his rebellion inevitably turns into a new Form' [Anders 2009, p. 35].

The *pupa*, as mentioned above, plays a special part as an instrument of Form. A subject dealt the *pupa* is frozen in its immaturity, vulnerable and susceptible to the influence of Form. So, above and beyond escaping Form, the subject needs to regain maturity. According to Gombrowicz, however, the trouble is that immaturity is a permanent and inherent part of subjectivity as such.

The face of a subject

If the *pupa* captures the subject in its immaturity, the transformative function of Form itself works differently. To be precise, we should speak more of the deforming than the moulding function of Form: under its impact, the face of a subject turns into a mug, which means it loses its authenticity; its expressions turn into posed and unnatural grimaces. If an individual seeks to become a mature and autonomous subject, he must 'save his Face'; the Face that is resistant to the *pupa* and the mug. The Face in Gombrowicz's anatomy can be identified as the subject; as such, this is not only a struggle to be authentic, but for being: 'Have pity on your faces, at least have pity on mine, **a face is not an object, a face is a subject, a subject, a subject!**' [Gombrowicz 2000, p. 63]. When salvaging one's face is at stake, one can use a mask to protect oneself. The face-pulling duel between Syphon and Kneadus shows how a once-destroyed face irrevocably changes its owner.

The Gombrowiczian Face plays the role of a place of identification, where all the inner qualities can be demonstrated. That is why one should be careful with it; once moulded into a mug, a Face can never regain its old features. On the other hand, a Face can also be altered through changes in character. This link between face and personality derives from physiognomics, a pseudo-scientific theory popularized by Johann Caspar Lavater, highly popular in the nineteenth century. Its main thesis was that a person's character, temperament and mentality is inscribed in his or her facial features and expressions. If we learn to read the traits and bone structures, we can ascertain whether a person is good or bad, lecherous or modest, mean or noble. These hypotheses, however absurd and scientifically unverifiable, inspired various more-or-less verifiable concepts such as pathognomy, phrenology, anthropological criminology and anthropometry [Skorupa 2013].

Gombrowicz's fierce adversary, literary critic Ignacy Fik, was the first to recognise the Face as the representation of a subject in *Ferdydurke*. He also read the conflict between the characters as an opposition between the biological and socio-cultural sides of human nature. The more one tries to resist the law of Form, the more one's Face is biologically determined and turns into a mug. One can only keep one's face by accepting socially ratified conventions of participation in the public sphere. It is the pure and virtuous Face that is recognized by Fik as most significant in this dichotomy: 'The human effort to rise above nature is the most glorious of things: to make his human will dictate its laws, even when it comes to himself, a particle thereof. To build clear and simple constructs and to read his work to structure the chaos through them. To create himself. This is difficult indeed. One errs a hundred times and grow discouraged. There may be a hundred failures. But there is no greater task. And there is no brighter joy than to see you have succeeded. Just as there is no more noble compassion than assisting another's efforts. Just as there is powerful suffering (the essence of cathartic tragedy) in witnessing the failure of sincere effort. One must seek man in his face (etymologically: *facere*!). There he is! Shall we flee from a mug? Indeed. Let's flee to our face. To our facial expressions!' [Fik 1938].[3]

'A Face' according to Fik, true to the idea of Gombrowicz, is a metaphor of 'who and how' a person is, a metaphor of his subjectivity. In these terms, a 'Face' is not a gift of fortune, a fact that exists 'in nature', but rather a task set before a person,

---

3 *Najcudniejszą rzeczą jest wysiłek człowieka, by być ponad naturą. By własną, ludzką wolą dyktować prawa nawet sobie, jej cząstce. By tworzyć konstrukcje jasne i proste i czytać w nich swą porządkującą chaos twórczość. Tworzyć siebie samego. Tak, to jest trudne. W tym można się sto razy pomylić i załamać. Sto razy może się nie udać. Ale przecież nie ma nic nad to. I nie ma piękniejszej radości, jak widzieć, że się udało. Jak nie ma szlachetniejszego współczucia: patrzeć na cudzy wysiłek, dopomóc. Jak nie ma mocniejszego cierpienia (istota oczyszczającej tragedii): widzieć klęskę szczerego wysiłku. — Człowieka trzeba szukać w jego twarzy (etymologicznie od: „tworzyć"!). Tam on jest. Uciekajmy od gęby? Owszem. Uciekajmy w twarz. W nasz wyraz!* [Fik 1938].

a project to be 'fulfilled'. Therefore, if one wants to be 'oneself', i.e. a person, one is condemned to 'artificiality', to being one's own 'creation'. The 'Face' always turns out to be a 'mask', an 'unnatural' and imitative construct on featureless and chaotic biological tissue [Werner 2009, pp. 128, 129].[4]

Every attempt at setting one's face against such rules and regulations elicits a twisted grimace, while it is the 'social face' that Fik admires. It is also the 'social face' that seems to save us from the mug, although Gombrowicz perceives it as a source of discontent and repression of the Self, which the restrictions of social rituals keep from fully developing. Moreover, the necessity of participating in such rituals (e.g. weddings, various family and religious celebrations, diplomacy or etiquette) results in a mug adhering to an individual. The condition for regaining one's 'true face', unmediated by 'culture' and the basis of an autonomous and sovereign subject, is impossible to achieve, according to Gombrowicz, but is based on the rejection of the 'social face' or rather, 'the social mug': 'Nothing but face, says a character in *Ferdydurke* who is looking for authentic countenance: the face that looks at me and the face it imposes on mine and the face I adopt in return and all the faces, mugs, grimaces, and permutations of phiz that pass between us. Just as that definitive paradox of Gombrowiczian space – "from a very near distance" – overrides any proprietary ideas about autonomous identity, so the face is synecdochic shorthand for consciousness and counter-moves by the face of another. One wears a face; one doesn't own it' [Newton 2004, p. 244].

If 'the face' can play the part of a place where the authentic, sovereign Self can be manifested, the mug is its falsification – a mask one wears to obey the rules of a game, when a situation requires codified and specific behaviour, words, and gestures. The mug is not only imposed on us by another, it is also something we wear, due to the subject's ability to mimic. But Gombrowicz broadens this definition of the 'cultural face' in literature. This textual signature that is inscribed in a text can be called the 'textual face' of a subject; it also uses various camouflages to avoid revealing itself; this is why Gombrowicz's attempt to find 'his own style' – original, personal and idiomatic – is so important to him as a writer. The challenge increases with the writer's awareness, so that such an attempt is nearly impossible – especially when it is the critic, professor or a 'cultural Auntie' who is the

4 *„Twarz" stanowi w argumentacji Fika – i zgodnie z duchem Gombrowiczowskiego pomysłu – metaforę tego, „kim i jak jest" człowiek, metaforę jego podmiotowości. W tym ujęciu „twarz" nie jest podarunkiem losu, faktem zastanym „w naturze", ale raczej narzuconym sobie zadaniem, projektem do „wypełnienia". A więc „twarz" okazuje się zawsze „maską", czymś „nienaturalnym", wtórnie nadbudowanym nad pozbawioną rysów i chaotyczną tkanką biologii.* [Werner 2009; pp. 128, 129]

judge. 'The author's struggle for his or her own authenticity becomes evident, but it is somehow transferred from the plot to a higher level of semantic units, it is exposed through applying particular conventions in juggling styles of utterances and the insistent avoidance of structural consistency in shaping a literary work' [Jarzębski 1984a, p. 20].[5]

Conclusion

The novels discussed above present only one aspect of the modern subject's struggle to regain a strong identity, to resist the fluidity of the modern world and the oppression of social rules and regulations. Both writers emphasise the performative dimension of subjectivity, which should actually be perceived as inter-subjective, as it is born and developed in relation to another subject, and perpetually susceptible to its influence. A Cartesian autonomy of mind, and therefore Self (which traps struggling subjects in an insurmountable paradox), is impossible to reach, however, as the modern individual sees the point of existence in building a strong identity. Those who are weak and susceptible to influence, prone to the manipulation of stronger subjects, are bound to intermingle, creating a new subject, with which modernists must learn to cope – mass society, a crowd consisting of identical, homogeneous subjects. In other words, this is a battle between the legacy of Romantic individuality and the modern homogenisation of society, which 'moulds' each person into its Form. It is a moan of protest against the world, where every face is identical, like a mask or a mug.

Bibliography

1. Anders, J., 2009. *Between Fire and Sleep: Essays in Modern Polish Poetry and Prose*, Yale University Press, New Haven.
2. Biasin, G-P., Gieri, M., 1999. *Luigi Pirandello: Contemporary Perspectives*, University of Toronto Press, Toronto.
3. Boyers, R., 1970. 'Gombrowicz and *Ferdydurke*: the Tyranny of Form', *The Centennial Review*, Vol. 14, No. 3, pp. 284–312.
4. Błoński, J., 1994. *Forma, śmiech i rzeczy ostateczne: studia o Gombrowiczu*, Wydawnictwo. Znak, Krakow.

---

5 *Zmagania autora o własną autentyczność stają się jawne, ale przenoszą się jakby na wyższy niż fabuła poziom jednostek semantycznych, ujawniają się mianowicie w zastosowaniu poszczególnych konwencji w żonglerce stylami wypowiedzi, w uporczywym uchylaniu się od wszelkiej formalnej konsekwencji w budowie dzieła.* [Jarzębski 1984a; 20]

5. Chwin, S., 2005. 'Gombrowicz i Forma polska', afterword in: Gombrowicz Witold, *Trans-Atlantyk*, Wydawnictwo Literackie, Krakow.
6. Fik, I., 1938. 'Miny trudne i miny łatwe. Uwagi na marginesie książki Witolda Gombrowicza *Ferdydurke*', *Nasz Wyraz*, No. 3.
7. Gombrowicz, W., 1997. *Ferdydurke*, eds. Jarzębski Jerzy, Zawadzki Andrzej, Wydawnictwo Literackie, Krakow.
8. Gombrowicz, W., 2000. *Ferdydurke*, trans. Borchardt Danuta, Yale University Press, New Haven, CT.
9. Gombrowicz, W., 1937. 'Aby uniknąć nieporozumienia', *Wiadomości Literackie*, No. 47.
10. Głowiński, M., 2002. *Gombrowicz i nadliteratura*, Wydawnictwo Literackie, Krakow.
11. Jarzębski, J., 1984. *Gra w Gombrowicza*, Wydawnictwo Literackie, Krakow.
12. Jarzębski, J., 1984a. *Powieść jako autokreacja*, Wydawnictwo Literackie, Krakow.
13. Jarzębski, J., 2010. *Witold Gombrowicz nasz współczesny: materiały międzynarodowej konferencji naukowej w stulecie urodzin pisarza: Uniwersytet Jagielloński, Kraków, 22–27 marca 2004*, Universitas, Krakow.
14. Kucharuk, S., 2011. *Pirandello i Szaniawski: Przyczynek do badań komparatystycznych*, Wydawnictwo Uniwersytetu Marii Curie-Skłodowskiej, Lublin.
15. Liszka, James, 1981. 'The Face: I and Other in Gombrowicz's *Ferdydurke*', *Philosophy and Literature*, Vol. 5, No. 1, pp. 62–72.
16. Lucente, G., 1984. 'Non conclude: Self-Consciousness and the Boundaries of Modernism in Pirandello's Narrative', *Criticism*, Vol. 26, No.1, pp. 21–47.
17. Markowski M. P., 2004. *Czarny nurt. Gombrowicz, świat, literatura*, Wyd. Literackie, Krakow.
18. Moroni, M. & Somigli, L., 2004. *Italian modernism: Italian Culture between Decadentism and Avant-garde*, University of Toronto Press, Toronto.
19. Newton, A. Z., 1998. *'Nothing but Face – To Hell with Philosophy*?: Witold Gombrowicz, Bruno Schulz and the Scandal of Human Countenance', *Literature and Ethical Criticism*, Vol. 32, No. 2, pp. 243–260.
20. Pirandello, L., 2011. *Jeden, nikt i sto tysięcy*, trans. Ugniewska Joanna, Państwowy Instytut Wydawniczy, Warsaw.
21. Skorupa, E., 2013. *Twarze - emocje - charaktery. Literacka przygoda z wiedzą o wyglądzie człowieka*, WUJ, Krakow.
22. Szymanowska, J., 2013. *Savinio, Pirandello i inni: szkice krytycznoliterackie*, Uniwersytet Warszawski, Warsaw.
23. Taylor, Ch., 1992. *Sources of Self: The Making of the Modern Identity*, Cambridge University Press.

24. Tellini, G., 2007. *The Invention of Modern Italian Literature: Strategies of Creative Imagination*, University of Toronto Press, Toronto.
25. *The Polish Review*, 2015. Vol. 60, No. 2.
26. Werner, M., 2009. *Wobec nihilizmu: Gombrowicz, Witkacy*, Sic!, Warsaw.
27. Zając, J., 2003. *Dwie koncepcje dramatu: D'Annunzio – Pirandello*, Księgarnia Akademicka, Krakow.
28. Łapiński, Z., 1997. *Ja, Ferdydurke*, Wydawnictwo Literackie, Krakow.

Beja Margitházi

Hybrid creatures and uncanny faces in digital cinema[1]

In the age of CGI, DFX, VFX and mocap, the creation and animation of digital characters is an everyday practice in both animation and live-action movies. Computer-generated humans in movies and video games are often trapped in the 'Uncanny Valley', as shown by some ambitious but unsuccessful movies (*Final Fantasy, The Polar Express, Beowulf, Adventures of Tintin*). These clearly prove that enhanced realism in artificial humanoids does not necessarily mean increased acceptance. To reframe this concept in the light of the latest advancements in CGI technology, I first review the original idea of the Uncanny Valley, and then, through the example of some photorealistic and caricatured computer-generated cinematic characters, I highlight motion capture technology in character-building. Finally, I argue that, although it has no computer-generated human characters, the latest installment of the rebooted *Planet of the Apes* saga (2011–2017) plays an important part in the long-running process of 'bridging the Uncanny Valley'.

Keywords: motion capture, Uncanny Valley, hybrid characters, digital visual effects, photorealism

Throughout the history of film, narrative cinematic images have often focused on human bodies and faces.[2] This practice has not changed even since the digital transition in film production. In the age of CGI, DFX, VFX and mocap,[3] the creation and animation of digital characters is an everyday practice in both animation and live-action movies, whether they are realistic human characters (like Benjamin Button) or fantasy creatures (like Gollum in *Lord of the Rings*). Although computer-generated humans in movies and video games are not yet indistinguishable from live actors, the oft-heard saying that lifelike digital humans are the 'Holy Grail' of computer animation strongly implies this intention. This search for the

1 During the research process, the author received the János Bolyai Research Scholarship of the Hungarian Academy of Sciences (project number BO/00140/11).

2 In my previous book I discussed the history and theory of cinematic face close-up in detail, see Margitházi 2008.

3 CGI stands for 'computer generated image'; DFX stands for 'digital effects'; VFX or 'visual effects' can also refer to computer generated or traditional, mechanical, special effects created during shooting, in front of the camera – or the combination of both effect types; mocap is an abbreviation for *motion capture,* referring to the digital recording of physical body movements.

Holy Grail often ends with being trapped in the 'Uncanny Valley', as we see in some ambitious but unsuccessful movies (*Final Fantasy, The Polar Express, Beowulf, Adventures of Tintin*), meaning that increased realism in artificial humanoids does not necessarily lead to increased acceptance. In order to investigate this problem I would first like to review the original concept of the Uncanny Valley, and then, through the example of some photorealistic and caricatured computer-generated cinematic characters, I will highlight the special importance of motion capture technology in character-building. Finally, I will argue that although it does not feature computer-generated human characters, the latest instalment of the rebooted *Planet of the Apes* saga plays an important role in the long-running process of 'bridging the Uncanny Valley'.

The uncanny Uncanny Valley

The concept of the Uncanny Valley (UV) began to be widespread in popular culture and over-cited in movie and video game reviews alongside the ascendance of computer animation in the creation and animation of human characters. Originally introduced in 1970 by Japanese robotics professor Masahiro Mori (Tokyo Institute of Technology), UV was initially used to illustrate the sudden change in human response to the humanoid movement and behaviour of anthropomorphic robots. According to Mori's hypothesis, people would first react more positively to robots having humanoid features, but the increasing affinity and empathy would suddenly change to rejection as the robot more closely resembled a human being. This sudden loss of empathy and the emerging sense of strangeness is illustrated in Mori's graph by a valley; he called it the 'Uncanny Valley', based on the sensation it generated. In his very short text of only three pages, Mori originally spoke of experiencing this phenomenon while working with industrial and toy robots; he illustrates it with the example of a prosthetic hand that initially looks very realistic, but after a handshake where we feel its coldness and artificiality, 'we lose our sense of affinity and the hand becomes uncanny' [Mori 2012, p. 99]. According to Mori, movement accentuates this impression (when the prosthetic hand begins to move, the sensation of eeriness intensifies), and a whole robot with a humanoid design intensifies the creepiness. Although today we often mention UV in relation to digital face design and animation, Mori initially used only the example of the face, in speaking of the plan for a highly sophisticated robot for the 1970 World Expo in Osaka, which had twenty-nine pairs of artificial muscles in the face in order to smile in a humanoid fashion. Here it was less the design of the face than the slowness of the smile that made the expression creepy [Mori 2012, p. 100]. Mori explained that this eerie sensation was an integral part of our self-preservation instinct, protecting us from proximal sources of danger (including corpses, different species, and other entities we can closely resemble). As Karl MacDormand recently summarises in his article, some possible explanations of the phenomenon employ concepts of *perceptual* processing, and other explanations remark on the importance of *cognitive* processing [MacDormand et al. 2009]. These explanations often stress the idea of fear and horror in identifying with cadaverous, mechanical creatures.

Despite its popularity, and the large number of UV-oriented scientific research over the last two decades, the concept of UV remains a credible, but unconfirmed hypothesis, lacking conclusive empirical evidence [Kätsyri et al. 2015, pp. 1–2]. A recent review article published by Finnish psychologists in March 2015 [Jari Kätsyri et al.] calls it, from a scientific point of view, a 'naïve hypothesis', because although it has motivated research in various fields beyond robotics and computer animation, including developmental psychology, neuroimaging, animal studies and philosophy, it still seems to be primarily an expressive metaphor for an emotional response that may be connected to the psychoanalytic concept of the 'uncanny' (or *unheimlich*) described by Freud and Ernst Jentsch in the early twentieth century [see: Tinwell 2014, pp. 2–6], and thus can be used to describe an uncomfortable effect, impression, or sensation of insecurity evoked by computer-generated characters. As we know from new experimental research, highly realistic computer-animated faces can already be produced [Kätsyri et al. 2015, p. 2], but contemporary computer animation techniques still tend to suffer from subtle imperfections related to rendering, lighting, surface materials, and movement dynamics, for example. In the last two decades, UV has become the standard term for near-photorealistic virtual humans that appear unintentionally eerie or creepy; it has been adopted to explain some films' lack of commercial success [see the works cited by Kätsyri et al. 2015, p. 2]. Low box-office returns as an expression of audience-opinion seem to function as clear-cut proof for the computer-generated characters of some respected movies falling into the UV, and it is generally speculated that this is due to audiences feeling a lack of empathy towards the characters. Thus, Mori's concept is taken seriously by filmmakers and game developers, as it suggests that unbecoming character design and movement can keep audiences from feeling emotionally engaged with their stories or games.

Motion-capture-based digital characters in cinema

In his book about digital visual effects in cinema, Stephen Prince singles out a new type of character in movies, one whose appearance is designed digitally, but whose movement is based on an actor's live performance, and transposed onto a computer with the help of motion capture technology. [Prince 2012, p. 117] It is indisputable that mocap-based digital characters have been a constant presence in Hollywood movies since the turn of the millennium; they can accompany live characters, in essentially live-action movies (e.g. *The Lord of the Rings: Two Towers* 2002, *Hulk* 2003, *King Kong* 2005, *The Curious Case of Benjamin Button* 2008, *Avatar* 2009, *Rise of the Planet of the Apes* 2011, *The Hobbit* 2012, *The Avengers* 2012, *Dawn of the Planet of the Apes* 2014), or as a character type in fully computer-animated movies (e.g. *Final Fantasy* 2001, *Polar Express* 2004, *Monster House* 2004/2006, *Beowulf* 2007, *A Christmas Carol* 2009, *Adventures of Tintin*, 2011, *Mars Needs Moms* 2011). According to Prince, the visual design of these characters exemplifies two broad stylistic objectives, *caricature* and *photorealism*. Caricature is present in the exaggerated, cartoonish, sometimes grotesque outlook of Gollum (*Lord of the Rings*), Grendel (*Beowulf*) or the Na'vi people (*Avatar*) – the design of these fictive, fantastical creatures intentionally differs from human beings. On the other hand, photorealism aims to create perceptually credible digital representations of human beings, as in the case of Benjamin Button, or the characters played by different actors in *Beowulf* or by Tom Hanks in *Polar Express*. The result of these efforts is ambiguous, as spectators and reviewers believe that some of these characters fall into the UV. In his original work, Mori advised designers of robots or artificial body parts that, in order to achieve a safe level of affinity, it is better to deliberately aim for a non-human design [Mori 2012, p. 100]; the same idea was formulated by Disney animators Frank Thomas and Ollie Johnston: 'If the shapes in the face and body can be caricatured just a little, the characters will be easier to animate and more convincing to the audience' [quoted by Prince 2012, p. 117]. This idea is well represented by the difference between the classic Disney animated figure of Snow White, rotoscoped after a live actor, and the more caricatured but lively dwarves, whose faces are far more expressive. Pixar films have a similar policy; in their first movies the main characters were objects or animals, and only in *The Incredibles* did the first human protagonists appear. Here Pixar used the same strategy of caricature to capture the character's inner personality, thus opting out on the quest for the Holy Grail of creating perfectly realistic human beings through digital animation.

In order to understand the effects produced by the most remarkable photorealistic and caricatured digital movie characters, we should have a clear view of the processes and the technologies used behind the scenes. In *Final Fantasy* (2001), for example, photorealist ambitions are evident. The characters' faces and appearances were first designed on computer, and the movements were then motion-captured with actors. The faces and hands were computer animated, and as in many other cases, the animators were inspired by their facial expressions and hand gestures.[4] In this way they were able to choose from fifty basic facial expressions, and they were able to control and manipulate the size of the pupils, the flaring of the nostrils, and the timing of the blinks. Despite all these efforts, they were unable to achieve expressive, photorealistic faces. As the crew later concluded, they lacked the technical capacity and resources to digitally model every exact detail, and they were probably using an inappropriate method to render the characters' skin to create a perceptually convincing human face.[5] According to visual effects animator Peter Plantec, the main character of *Final Fantasy*, Aki Ross, comes across as 'a cartoon character masquerading as a human. As she moves, our minds pick up on the incorrectness. And as we focus on her eyes, mouth, skin and hair, they destroy the illusion of reality' [Plantec 2007].

In Robert Zemeckis' *Beowulf* (2007), the texturing of the characters' hair, clothes and other surfaces was rendered in microscopic detail, producing an almost photographic quality, but the subtleties of the expressions of the actors (Anthony Hopkins, John Malkovich, Ray Winstone, Robin Wright Penn) were less than successful. The performances seem mechanical, they lack life and presence despite the visible similarities between the animated characters and the famous actors and the detailed surface texturing of their skin. At the same time, working with motion capture created conditions that most resembled the theatre: the set was covered with over two hundred mocap cameras, which meant actors were able to perform very long takes, entire scenes running through from the beginning to end. Despite all of this, a great deal of facial information from their performances could not be rendered into the final screen characters, and many viewers have found them weird and creepy.[6] The mixture of live action and painterly digital animation creates a dissonant and contradictory impression, as in Zemeckis' other mocap films (*Polar Express*, *Beowulf*, *Mars Needs Moms*). Audiences of *Beowulf* were left with the strange feeling that the grotesque monster, Grendel, was the most believable and affecting character in the whole film [Prince 2012, p. 125].

The ghostliness of the characters in *Final Fantasy* and *Beowulf* aptly illustrates the difficulties of photorealistic digital reproduction of moving faces. More than other parts of the body, the face contains muscles that move in complex, subtle ways and generate a great range of expressions; moreover, these muscles are not

4 On the work process see: http://en.wikipedia.org/wiki/Final_Fantasy:_The_Spirits_Within (accessed: 15.05.2015).

5 For a summation of the animators' opinions see Prince 2012, pp. 120–121.

6 For technical description and details on viewers responses see http://en.wikipedia.org/wiki/Beowulf_%282007_film%29 (accessed: 13.05.2015).

attached to bones, but from skin to skin. As Prince sums up, facial muscles are unable to produce 'vertex data' that is essential for motion capture, and is easily produced by other body parts (the hands, feet, spine), whose movements can be arranged around different vectors [Prince 2012, p. 123]. The complexity of facial movement is hard to cover by numerous markers (as were used on Tom Hanks during the shooting of *The Polar Express*), and the eyes cannot be motion captured at all. The reproduction of eye movement (blinking, direction of gaze etc.) is left up to animators, but the smallest mistake can lead to 'dead eye'-effect that can suddenly turn any photorealistic character into a zombie-like creature [Tinwell 2014, p. 18].

Many critics and VFX specialists agree that the first perceptually realistic and believable computer-generated photorealistic human character achieved through combining live action performance and digital animation in the protagonist of *The Curious Case of Benjamin Button*.[7] In the live action film we see another form of hybridity in the case of the character played by Brad Pitt: Benjamin is living backwards, he is born old and dies as a baby. In the first thirty years (from his eighties to his late sixties) Benjamin is played by three different actors, whose bodies were composited with the head of Pitt, made to look digitally older. Later Pitt appears physically as full-body actor, wearing digitally refined prosthetic age make-up until the scenes when the character's age and the actor's age correspond: then we see Pitt without any digital effects. Later, as he aged back during the plot, Pitt was digitally made younger. Despite the different digital and prosthetic layers, Pitt's facial expression is much more vivid than the other examples mentioned earlier. David Fincher's crew used another kind of motion capture technology, one based on *Mova* phosphorescent make-up worn by the actor. As this make-up forms a continuous layer on the face, even the micro-expressions performed by Pitt could be transposed to his digitally aged or youthened faces.

Among other emphatic and expressive mocap-based digital characters we find such fantastical creatures as Gollum or the Na'vi from planet Pandora. As they represent non-existent, fictitious creatures, instead of referential photorealism, it seems more suitable to speak of the kind of 'perceptual realism' Stephen Prince introduces in the case of the dinosaurs in *Jurassic Park*. This refers to the digital creation of a reality structurally similar to the viewer's everyday, three-dimensional, audiovisual experience [Prince 1996, p. 32].

Andy Serkis' mocapped performance served as the basis for the computer-generated character of Gollum, but the animators only used his facial expressions

7 The creation of the character is described in detail by Flueckiger, 2011, pp. 7–15.

as a point of departure in creating the character's final facial movements. The most expressive part of Gollum's body was his huge eyes, which were able to enlarge the monomaniacal creature's obsession and inner agony to possess the Ring.[8] Gollum's exaggerated, but emotionally affective character is part of the same caricaturial tendency as the Na'vi people of James Cameron's *Avatar*.[9] The *Simulcam* camera and computer system helped Cameron obtain composite CGI and live-action footage while shooting. The mini cameras mounted to a headpiece worn by the actors registered their facial performances (*performance capture*), which were later refined and animated with the help of Pail Ekman's Facial Action Coding System (FACS), developed by the professor in the 1970s to capture the main facial expressions through muscle movement.[10] The special Na'vi physiognomy of the main protagonists was created on the basis of the actors impersonating them, with special attention to eye movement. In this way digital animation integrated the presence, movement, physiognomy and voice of the live actors into CGI characters [Prince 2012, p. 135].

In all these films mocap-based digital characters are visual hybrids, created by compositing different image types and techniques (live-action shots, computer animation). The photorealism and caricature described by Prince pertain to different species: photorealistic mocap-based and digitally designed human characters often arrive in the Uncanny Valley (*Final Fantasy*, *Beowulf*), and they can only cross it when adequate technology meets a strange character (Benjamin Button). With fantasy creatures (Gollum, the Na'vi), caricatured design that blends animation and live action results in 'perceptually realistic' characters, which indicates that digital technology is, at present, mainly successful in creating these kinds of creatures. The concept of the Uncanny Valley was originally proposed for human imitations – the frog-like Gollum and the blue Na'vi creatures are not humans, but by their diegetic ontology they happen to be hybrids, humanoids in different degree and manner: the schizophrenic Gollum was once a hobbit, Jake and Dr. Grace are humans who later become avatars, and Jake himself is ultimately transformed into a Na'vi.

Intelligent, uncanny apes

From the successful adaptation of the original satirical French novel in 1968, *Planet of the Apes* has gradually become a science-fiction media franchise consisting of movies, books, television series and other media, about a world where

8 For a detailed description see Prince 2012, p. 133–136.

9 Flueckiger offers a general review of the process: 2011, pp. 15–26.

10 On the FACS system, see an earlier work by Margitházi 2008, p. 74.

humans and intelligent apes are fighting for control. Special effects have always been involved in its screen adaptations, documenting the state of the art in the respective age. The latex masks of the original 1968 movie were considered revolutionary at the time, but they cannot be compared to the digital visual effects of the latest *Apes* reboot sequel, which began with 2011's *Rise of the Planet of the Apes* and continued with the *Dawn of the Planet of the Apes* in 2014. *Dawn* picks up ten years after the events in *Rise*: the 'Simian Flu', a man-made virus, has got out of a lab, wiped out most of humanity, and left hyper-intelligent apes, who have built a pacifist society in the woods near San Francisco. In these two films human actors 'play' monkeys, as in the earlier parts, but the prosthetic masks have been replaced with digital animation in such a convincing fashion that *Rise* (2011) and recently *Dawn* (2014), seem to be landmarks in the history of cinematic computer-generated character building, even if the films themselves regurgitate typical Hollywood formulae and clichés.

Weta Digital (who worked earlier on *King Kong* and *The Lord of the Rings* and *Hobbit* trilogies) planned to design realistic-looking digital apes, later adding performance factors from motion-capture shots of the actors. The characters needed to recognisably resemble apes, yet 'behave unlike anything ever seen before' [Letteri, in Powell, 2014]. In order to create believable and emotionally engaging characters, the VFX crew had to concentrate on the visual design of the apes, the performances of the actors and the visual integration of the two. The digital animators first studied ape physiology in order to build believable skeletons, muscle systems with layers of tissue, skin and fur. *Dawn of the Planet of the Apes* was also shot under new conditions: before then, motion capture technology had facilitated shooting on a controlled studio set, with a green screen background, but here it was possible to shoot on location, in the green of woods, inside the city and in various buildings.[11] During post-production, the animators worked to integrate the mocapped performances of human emotions and expressions with the ape physiology to create believable chimp behaviour. [12]

According to many critics and VFX specialists, *Dawn* succeeded in creating the remarkably photorealistic appearance, movement and behaviour of the digitally composited creatures. Compared to earlier hybrid characters, the hyper-intelligent

11 On shooting and post-production work, see a longer interview with the VFX specialists: http://www.artofvfx.com/?p=8794 (accessed: 17 March 2015) and Correy Powell's summary (Powell 2014).

12 This final and complex matching and integrating phase of the work was described by Joe Letteri as a 'mix of science and art' [Powell 2014], since the VFX team had to strive for both physical accuracy and artistic interpretation.

apes are semi-fantastical and semi-realistic creatures, as they are specially endowed members of an existing species. These apes are anthropoids, even humanoids; they behave, move, feel, think and even talk like humans do. The dramatic characters (the wise leader Caesar, his son Blue Eyes, the intelligent but hot-tempered Koba etc.), the Shakespearean conflicts of betrayal, revenge, war and forgiveness in the narrative, as well as the animal morality and ethics joined with the photorealistic representation of the hyper-intelligent apes, together make a powerful overall impression. Besides the successful integration of motion capture, actor performance and ape physiology, many other cinematographic and diegetic narrative elements accentuate and strengthen the film's photorealistic effect.

Interaction. Humans and apes often appear together, in a natural landscape, in the same space. In these scenes, the CGI apes interact with live actors in an active and spatially convincing way. In some key moments – such as the first meeting of humans and apes in the woods, or when the apes first come to the city, or Koba's confrontation with the two armed men – human presence even involves reactions of surprise and shock toward the talking and hyper-intelligent apes, thus modelling audiences' response to the believable computer-generated characters.

Speech. By presenting speaking apes, the plot of the film accelerates the evolution of chimps toward becoming intelligent beings. Among themselves the apes use sign language, but as the humans appear, they begin to use dialogue. We see their struggle to articulate themselves, but at the same time, the learning process of speaking and using language is a means of self-expression in dramatic situations.

Faces and eyes. *Dawn* contains a lot of close-ups on the apes' faces that contribute to the spectacular effect of the film, showing high-precision CGI details. The close-ups often accentuate the vibrant eyes of the simians, in contrast with the 'dead eyes' of earlier uncanny characters. The eyes of the apes here are emblematic visual and diegetic motifs, as the first visible sign of the effect of genetic serum causing Simian Flu is the change in the colour of the ape's eyes, from blue to green.[13] The extreme close-up on Caesar's eyes frames the two movies: it was the first shot of *Rise*, and it is the final image of *Dawn,* and it is repeated several times during the film. Carefully computer generated from facial capture images of actors, the apes' eyes – like human eyes – become a form of identity and expression: in these close-ups they are more visibly able to look, focus, and express intelligence, refined feelings and thoughts.

---

13 The name 'Blue Eyes' was given to Caesar's son because he represents the first generation that was born without the chemical that changed the other apes' eyes green when they lived in the laboratory.

Metaphorical hybridity. Compared to Gollum and the Na'vi, these intelligent apes are literally and explicitly humanoids. The animals appearing in the film are not real apes, monkeys or gorillas, and are not fully animated CGI creatures – technically they are analogue/digital hybrids. The digital incorporation of human presence and actors' performances into the characters is practically and technologically modelled, mirroring the effects of the serum, the human factor in the apes. The process of genetic modification in the plot (the apes' transformation into intelligent humanoid apes), is repeated, modelled through the technical creation of these characters, as the human factors appear and become integrated into their behaviour. Unlike the caricatured representational style of the Na'vi and Gollum, the hybrid apes are both convincingly photorealistic in their appearance and perceptually realistic in their behaviour – no wonder that, as hyper-intelligent creatures and dramatic characters, they seem to be *uncannily* realistic.

Motion capture technology might be considered a late successor of the rotoscope animation technology used in the 1910s by the Fleischer brothers, as the two have a great many features in common. Mocap became popular in shooting feature films in the nineties, and was developed to reproduce facial movements in the 2000s. Although it was often used for computer-generated fantastical creatures or human characters represented in photorealistic or caricature style, it was rarely successful in creating believable humans who did not fall into the Uncanny Valley. All these efforts, on the other hand, make it clear that motion capture is more than a simple technical process of shooting and digital post-production: it profoundly affects the spectator's emotional response to computer-generated characters, as these creatures are somehow familiar, human in their movements, gestures and behaviour, but – due to the imperfections that still exist in digital reproduction – their design and appearance remains alienating and artificial. This special in-betweenness characteristic of the present state of the art seems tailor-made for such kinds of humanoid, fantastical or animal characters whose hybridity is justified and confirmed by the narrative and the diegetic world of the movie in question.

Bibliography

1. Flueckiger, B., 2011. 'Computer-Generated Characters in Avatar and Benjamin Button', in Segeberg H. (ed.), *Digitalität und Kino*. Munich: Fink. English translation: http://www.zauberklang.ch/AvatarButtonFlueckiger.pdf (accessed: 10.10. 2015).
2. Kätsiry, J. et al., 2015. 'A review of empirical evidence on different Uncanny Valley hypotheses: support for perceptual mismatch as one road to the valley

of eeriness', *Front. Psychol.* 6:390. Available at: www.frontiersin.org (accessed: 01.03.2015)

3. MacDormand, K. F. et al., 2009. 'Too Real For Comfort? Uncanny Responses to Computer Generated Faces', *Computers in Human Behavior*, 25(3), pp. 695–710.
4. Margithazi, B., 2008. *Az arc mozija. Közelkép és filmstílus* [The Cinema of the Face: Close-up and Film Style], Kolozsvár: Koinónia.
5. Mori M., MacDormand K. F., Kageki N., 2012. 'The Uncanny Valley [from the field]', *Robotics & Automation Magazine*, IEEE, 19(2), pp. 98–100.
6. Ng, J., 2012. 'Seeing Movement. On Motion Capture Animation and James Cameron's Avatar', *Animation* 7:3, pp. 273–286.
7. Plantec, P., 2007. *Crossing the Great Uncanny Valley.* Animation Word Network. Available at: http://www.awn.com/vfxworld/crossing-great-uncanny-valley (accessed: 12.06.2015)
8. Prince, S., 1996. 'True Lies: Perceptual Realism, Digital Images and Film Theory', *Film Quarterly* 49.3, pp. 27–37.
9. Prince, S., 2012. *Digital Visual Effects in Cinema: The Seduction of Reality*, Rutgers University Press.
10. Powell, C. S., 2014. 'Digital Science behind the Dawn of the Planet of the Apes', Available at: http://blogs.discovermagazine.com/outthere/2014/07/12/mind-blowing-digital-science-behind-new-planet-apes/#.VQ1IkI6vMyw (accessed: 15.06.2015)
11. Tinwell, A., 2014. *The Uncanny Valley in Games and Animation,* CRC Press.

Olga Szmidt

The videotaped confession: Authenticity and the suicidal spectacle of the face. A study of a fragment of the *Homeland* TV series[1]

Authenticity and the body is are important issues in contemporary culture, and pop culture and new television series are no exceptions. By culturally, philosophically and ethically examining this important issue, this article discusses how authenticity is constructed through the human face and body. It focuses on some aspects of confessions by the main characters of the *Homeland* television series. The most important aspects are: how the body and the face are presented in acts of videotaped confession and the suicidal act of confession. This programme clearly shows authenticity as a social convention and a staged act. Depicting and preparing authenticity in a video recording is linked with blackmail in modern society and culture. Another issue considered in the article is how the body can be manipulated by the media and the individual to stage authenticity in a 'truth tape'.

Keywords: Homeland, television series, authenticity, videotaped confession, truth tape

Research on the face and its meaning in culture has revealed a complex picture. On the one hand, the face has been a philosophical and artistic obsession for such great European directors as Ingmar Bergman or Andrei Tarkovsky. On the other, contemporaneously, it has been a handy tool for advertisements and rather unsophisticated commercial campaigns. The role of the face in popular culture seems to be located in-between. Correspondingly, popular TV series explore the ambiguity of the idea of authenticity in contemporary society, politics and culture.

This article will deal with the notion of authenticity, which has become a great issue for both contemporary cinema and TV series. Twenty-first century problems like gender inequality, the cynical world of politics, various addictions or the financial crisis and its consequences seem to be more attractive for television than the deep-rooted idea of authenticity. Moreover, these problems arise in the same productions, such as *Breaking Bad* (AMC), *House of Cards* (Netflix) or *Homeland* (Showtime). This last example provides an opportunity to explore the idea of authenticity as a challenge for a TV series. In this paper I would like to focus on a fragment of the *Homeland* TV series which contains a videotaped confession by the lead male character. Specifically, my interpretation will address

1 The article was written in 2014, before the release of the 4th season of *Homeland*.

several issues, such as: the construction of the face and body in the videotaped confession, the multiplication of the self in this performance, the suicide confession as a suspended death or authenticity which turns into blackmail. The article will provide a general philosophical context for approaching authenticity, as well as the framework of quality television. The juxtaposition of these two points of view might help us to understand the ambiguity of authenticity in contemporary television.

All this leads us to the following question: What happened in culture in recent decades so that we must revisit this conservative idea? The explosion of authentic shows and authentic art made by authentic people is not and cannot be accidental. However, it is not without precedent. As Lionel Trilling writes:

> The word [sincerity – O. S.] as we now use it refers primarily to a congruence between avowal and actual feeling. Is it really possible, does it make sense, to say that the value put upon this congruence became, at a given moment in history, a new element of the moral life? Surely it is as old as speech and gesture? But I subdue this scepticism by reflecting that the word cannot be applied to a person without regard to his cultural circumstances. For example, we cannot say of the patriarch Abraham that he was a sincere man. That statement must seem only comical. The sincerity of Achilles or Beowulf cannot be discussed: they neither have nor lack sincerity. But if we ask whether young Werther is really as sincere as he intends to be, or which of the two Dashwood sisters, Elinor or Marianne, is thought by Jane Austen to be the more truly sincere, we can confidently expect a serious response in the form of opinions on both sides of the question. [Trilling 1972, pp. 2–3]

The gist of this fragment covers the socio-cultural circumstances which not only let us speak of authenticity, but also define rules prescribing how we can use it or even inquire into it. I see authenticity as an exclusively modern idea. This means that we cannot elaborate on the *Confessions* of Saint Augustine as an authentic autobiography – this complex issue begins with Jean-Jacques Rousseau's work. Crucially, authenticity is an idea that could not have been invented before modernity. In other words, authenticity is a fundamental part of modernity, just as modernity built the foundations for the emergence of the idea of authenticity. Therefore, when I use this term I always mean one of the most important obsessions and hopes of modernity, one that is pivotal for philosophy, literature, art and ultimately, popular culture [Warchala 2006].

As Trilling remarks,

> [...] that the word has become part of the moral slang of our day points to the peculiar nature of our fallen condition, our anxiety over the credibility of existence and of individual existences. [Trilling 1972, p. 93]

What surprises me the most is that today's moral slang finds a number of guileless enthusiasts who give speeches rife with such slogans as 'Be yourself no matter what they say", 'I am strong because I know who I am', 'If you can't be yourself, you're nobody' or 'Be your own inspiration'. By this rhetoric, inner authenticity combats artificiality and creation, the social evils which are trying to kill our strong and untarnished souls. But what is supposed to be the origin of this power? It comes, of course, from our natural condition and pure hearts. This seems to be a controversial point of view, to put it mildly, especially when we consider how little contemporary culture has in common with Rousseau's world. Trilling observes:

> From Rousseau we learned that what destroys our authenticity is society – our sentiment of being depends upon the opinion of other people. The ideal of authentic personal being stands at the very centre of Rousseau's thought. Yet I think that its presence there, however forcible it may have seemed to Rousseau's contemporaries, is rather too abstract, or too moderate, to command the modern imagination. […] Nowadays our sense of what authenticity means involves a degree of rough concreteness or of extremity which Rousseau, with his abiding commitment to an ideal of patrician civility […]. [Trilling 1972, pp. 93–94]

What Rousseau left us with, no doubt, is a dream which has never come true, or merely an utterly outdated literary fiction. I will have to skip further reflections on the history of authenticity and the reasons why we invent and use so many disparate definitions for the same term. I will stick to my findings on contemporary culture.

It is believed that modern media has a significant impact on how we perceive everyday experience. I do not mean just the omnipresence of television, music, the Internet etc., but primarily the constant mediation of experience through texts, videos, other voices or photography. They form our experience, subjectivity, relations with others, memory and our analytical and critical methods. This means that we can be as complicated as the most sophisticated literature – hard to interpret, understand or even to depict. What I am briefly outlining is one of the most rudimentary issues – the modern self or, one might say, identity. 'Self trouble' has had a huge and underestimated impact on the idea of authenticity, but we can agree that these two topics – the self and authenticity – have much in common. That is why in this case study I will use them as two ways of approaching the same problem.

As my focus of interest I have picked a powerful social force, an American TV drama called *Homeland,* which complicates the problem of authenticity to a much higher degree than many academic studies on this topic. However, we first need to outline the horizon of contemporary TV and the socio-cultural context of 'quality television'. *Homeland* (original release: 2011-present) is part of a new wave of television, labelled 'quality'. This was developed in the 1990s by American cable TV networks (especially HBO). It created not only a new style of television production and aesthetics, but also a different relationship between the viewer and the television drama, including intellectual and emotional investment. Here are a few general remarks on quality TV. Robert Thompson writes:

> 'Quality TV' had become a super-genre, a formula unto itself. By the turn of the century, quality was busting out all over the networks. [Thompson 2011]

And:

> The precise definition of 'quality TV' was elusive right from the start, though we knew it when we saw it. These shows were generic mongrels, often scrambling and recombining traditional TV formulas in unexpected ways; they had literary and cinematic ambitions beyond what we had seen before, and they employed complex and sophisticated serialized narratives and inter-series 'mythologies'. [Thompson 2011]

The definitions of this type of TV series suggest that it is more accurate to isolate this genre not by comparing it to other TV shows, but to other media. The presupposition seems quite clear – there are some shows whose forerunners are not in television, but in cinema, literature or theatre. *Six Feet Under, The Sopranos* or *Twin Peaks* are perhaps the best examples of this artistic strategy.

Jane Feuer observes:

> The claim to being 'not TV' is made by claiming to be something else: namely, art. cinema or modernist theatre. [...] The fact that the show [*The West Wing*] claims to be totally unique and original only reinforces my point; quality drama always claims to be original in relation to the regular TV norms of its era. [Feuer 2011]

Distancing themselves from traditional TV, quality series aspire to create a new space where conventional television rules do not apply and where originality and individuality are the most desired qualities. After all, quality series are often hybrid genres, created for a single production. Brett Martin claims in the book *Difficult Men. Behind the Scenes of a Creative Revolution: From 'The Sopranos' and 'The Wire'* that the protagonists also changed:

> In keeping with their protagonists, this new generation of shows would feature stories far more ambiguous and complicated than anything that television, always concerned with pleasing the widest possible audience and group of advertisers, had ever seen. They would be narratively ruthless: brooking no quarter for which might be the audience's favorite characters, offering little in the way of catharsis or the easy resolution in which television traditionally traded. [Martin 2013, p. 5]

Moreover, quality series have great ambitions to raise and diagnose crucial problems of contemporary society; this did not evolve until the current, third Golden

Age of Television (2000s-present). Entertainment is not the only aim; it seems equally important to interpret ideas which haunt society and individual lives. It is generally agreed that this kind of TV series provides a space for tackling issues which remain unresolved, but which play a defining role for the human condition in the twenty-first century. This is probably why authenticity found Nicholas Brody (*Homeland*), Walter White (*Breaking Bad*) and Lena Dunham (*Girls*) and forced them to get involved in a process which we could call 'redefining authenticity in contemporary culture'. This change is happening literally before our eyes. To paraphrase a popular HBO slogan: it's not TV, it's TV of authenticity.

The first season of *Homeland* was released in 2011 and the series is still being produced. The main character is Nicholas Brody, a USMC Sergeant who was taken hostage by Al-Qaeda for eight years. *Homeland* begins when he is on his way home. The drama depicts his life after the return, mingled with some retrospective scenes from Afghanistan. Brody is being watched by Carrie Mathison, who has set up an illegal surveillance operation in his house because she suspects that Brody was 'turned'. A video is part of the first season finale. We have already seen preparations for a terrorist attack on the Vice President, who is responsible for a raid on a school in Damascus that caused the deaths of eighty-two children, including the son of a fictional[2] Al-Qaeda terrorist, Abu Nazir – Brody's guru.

After this recording Brody changes into civilian clothes and leaves a memory card with a tape at a drop-off point. We can isolate two problems associated with this video, which is presented in a doubly-mediated manner. The whole drama is 'video', but his recording includes technical aspects of its own creation. We not only watch the confession tape as such, but are also witness to the fact that it is a tape. We see the characteristic camera frame and the REC symbol which signals that the recording is taking place. Why can we not have this confession without these mediating signs? In my opinion, these features primarily work as an act of isolating this recording from the episode and the drama as a whole. Secondly, it is a message that we are watching something that is being recorded on memory cards (it is like Chekhov's gun) in *Homeland*'s world. Finally – and most significantly – it is a signal that we are seeing an event of a profoundly conventional nature. Videotaped confessions have a long tradition which is immediately recognisable by the viewer.[3] If we see the top half of a man wearing a military uniform and with a certain facial expression, we expect a serious speech with important explanations

2 A fictional character inspired by real figure, Abu Nasir, who died in 2006.

3 This is also a popular strategy in quality series. Videotaped confessions are used as part of the story in *House of Cards* and *Breaking Bad*, among others.

or a confession of plans and decisions. It is significant how he dresses himself to make a convincing confession, while leaving his body in the background to highlight the face.

We should note that the character's gaze is fixed on the lens, which makes the viewer unable to avoid eye contact. His eyes are still, and keep intensifying the impression that Brody is determined not only to attack the Vice President, but also to record this confession. He tells the truth, looking into our eyes, and needs no technical interference (cuts, sound, movement). This raw aesthetic contains a message – I am here just to tell the truth and nothing else matters, not even myself or my body.

The *Homeland* case is a very striking one. The man is evidently dressed up for this event – he is wearing a Marine uniform, which is easy to recognise, but also signals a serious and official occasion (Marines wear uniforms in every public situation). We have left the domain of the private. Furthermore, it is a sign that 'being a Marine' is crucial to this confession and the actions mentioned in this speech. The tape is not for private or family viewing. In contrast to Walter White's taped confession in the first episode of *Breaking Bad*, it is hard to determine the addressee of Brody's explanation – he does not reveal this person, but creates a general 'you' who does not know the truth about him and his future actions.

However, this is not a traditional sequence of gestures and statements in a confession. As J. M. Coetzee writes:

> Confession is one element in a sequence of transgression, confession, penitence and absolution. Absolution means the end of the episode, the closing of the chapter, liberation from the oppression of the memory. Absolution in this sense is therefore the indispensable goal of all confession, sacramental or secular. [Coetzee 1985, p. 194]

Brody focuses on defining who he is and on the essential meaning of his plans: their rationale and whom they are directed against. We can guess that the addressee is public opinion in general, which does not exclude his family. It is also worth noting that Brody does not record two confessions, one for the public and one for his relatives. In this case, the confession is almost universal – he talks about the future and the past, but we do not need to have extensive knowledge of him or the political situation to understand the confession. All we have to know is that the USA is in conflict with terrorists. Apart from introducing himself (providing personal details, concerning the Marines and his family situation), Brody focuses on confessing why he intends to commit a mysterious act before anybody could watch this video. This means that the videotaped confession has no informational value, it is merely explanatory.

After the final explanation and providing an interpretation of what happened, the performance concludes. In other words, he elaborates on his motivation and his function in the terrorist system. His declaration carries the final authenticity he stages. The truth being confessed by the self is tantamount to the death of the self.[4] This is the way Brody has chosen by deciding to record the confession.

Essentially, recordings of this sort cannot be seen as equivalents of literary confessions [Gunn 1982, p. 5], primarily due to the stylisation of the body and of the face that literature precludes. There is no way, however, to tape a confession without these 'operations', which make the human face recognisable as the key element of human identity – an element which might be a part of his/her creation. In this video, Brody is motionless (like his camera) and does not gesture. Apart from some quavering (when he talks about his family), his voice is firm, strong and determined to express the whole truth. It also tells us that this man is not only ready to commit crimes, but also to interpret them. Brody is the master of meaning, which can be seen as a double victory.

His speech appears to be pre-prepared and deliberated, but not learned by heart, which strengthens the rhetorical effect of the confession. In many ways the situation differs from recordings of witnesses described by James E. Young's article [Young 2004, pp. 245–265], on the relationship between the speaker, the director and the camera. Brody is the one and only actor in this fatal show. He is confessor, a director and the creator of an empty space for the audience. His face and body are inert, he is the only one who reacts to his words. Nobody is coercing him except for himself or some imagined others. Only the self verifies what the self says – there is no interviewer to change his stream of words, ask deeper questions or convince him to say more [Young 2004, pp. 245–265]. This is another reason why confession tapes are taken so seriously and have such a shattering effect – there is no one to blame for the form of confession. There is no one to blame for manipulation or constraints, to own up to something that one did not do.[5] J. M. Coetzee notes in this context:

> Oswald Spengler, quoting Goethe's lament over the end of auricular confession brought about by Protestantism, suggests that it was inevitable that after the Reformation the confessional impulse should find an outlet in the arts, but also that, in the absence of a confessor, it is inevitable that such confession should tend to be 'unbounded'. [Coetzee 1985, p. 195]

4 This is not equivalent to what is said in postmodernist manifestos like *What is an Author?* by Michel Foucault or even *The Death of the Author* by Roland Barthes.

5 Of course, we know that the self is not as innocent to itself as to the eye of camera, but that is another issue.

Confession in a TV series is not like Saint Augustine's model and has no power to save Brody or his body from public attack. In this case, it is his own end; the explanation of his death is provided through the overriding idea. This video can be seen by the public only upon the death of the subject. Thus, there is a complementary mechanism – the death of the subject must follow, which I would call the escalation of subjectivity. What is more, Brody uses a particular phrase: *that's why I want to explain myself.* Obviously, explaining the self is the last thing that others seek to hear from him. *I am not there so I can say who I was and why I am gone.* The subject of the bombing projected in this short video is a dead subject – unable to communicate, leaving only a posthumous message. On the whole, the question of whether a confession can be true remains unresolved. If we focus on 'staged authenticity' [MacCannell 1976] (applying Dean MacCannell's term to a different context), we might better grasp the consequences of this videotaped performance for Brody's self.

Here the main motif of the drama – apart from American patriotism and the world division introduced in the initial conflict (the democratic West versus violent terrorists) – concerns the idea of truth, the possibility of using it as blackmail and revealing the true face to be yet another mask.[6] The truth is built through thousands of interrogations, wire-tapping and surveillance, repeated confessions to family (Brody-Jessica, Brody-Dana), lovers (Brody-Carrie, Carrie-Brody) and friends (Carrie-Saul). Yet since we are dealing with a show of a slightly more complex structure, often undermining the credibility of its characters (the leader is Carrie Mathison, a character with bipolar disorder and psychotic visions), the plan cannot succeed. The key to understanding this drama is recognising how we perceive Carrie's consciousness and our trust in her judgements. We cannot be sure what is real and what is the truth when we see the world through her eyes and interpretations. If she is trustworthy, we know a lot, but if she is not, we know absolutely nothing because of her central role in drama.

In the *Homeland* universe, this kind of tape has to function according to special rules. The tape is justified by its uniqueness (there are no other videos like it) and – paradoxically – its conventionality. During all of the forty-eight episodes, this is the only piece whose legitimacy is not disputed. Nobody denies its authenticity or the confession's basis in reality. No one even suggests the possibility of another interpretation. The thought that Brody could have been forced to deliver this

6 The idea of revealing a mask instead of the real face is quite common, i.e. Hans Belting, *Faces. Historia twarzy,* trans. Tadeusz Zatorski, Wydawnictwo Słowo/Obraz Terytoria, Gdańsk 2014, p. 250.

speech does not arise. Interestingly, Brody himself says that he supposes that his declarations might be seen as resulting from his being broken or brainwashed; yet, at the same time, he rejects the possibility of his unreliability. The convention of the video, his way of speaking and his appearance, seem to predetermine the public verdict: Brody told the truth – his body, voice and face certify it. Apparently, convention differentiates between textual and video confessions – we would probably be ready to expect falsification in a suicide letter, but we forget to be critical in front of the TV. The screen seems to eliminate the possibility that his confession is untrue. A 'truth tape' has to be true – otherwise our fragile world will collapse.

As expected, Brody's plan is not carried out – he changed his mind and decided to save his life and others'. Therefore, we could claim that his confession is no longer valid. The memory card turns out to have vanished, however, so the confession is used many times in further episodes. I want to focus on one of these. During the interrogation after the 'unsuccessful bombing', Brody deceives Peter Quinn, a CIA agent. He lies about his experiences in prison, collaboration with terrorists and bombing plans. Notably, his declarations are called into question by flashbacks from Afghanistan (often featured in the drama) – we see a story unlike the one that Brody invents. It is a very simple trick – the recollections are filmed in different colours, motion and sound from all the rest of action, creating the impression that they come from a world external to the show and that Brody's memory (despite his extreme experiences) does not lie. But what turns out to be a breakthrough in the whole interrogation and an ordeal for the subject is not a product of his memory, but his confession. For the subject this is the most extreme situation, equivalent to acknowledging his own death as a confessor. When the recording is shown to Brody, it makes the subject nothing less than a spectator to his own death as a subject.

Furthermore, Brody comes face to face with himself speaking; he confronts the time when he will no longer be among the living. The situation is nerve-wracking, which adds to the overall feel of the show, and it is difficult to predict the subject's response or reaction. If the main emotion were shame, we would consider its effects on the self. An interesting approach to this topic is proposed by J. A. Deonna and F. Teroni in an article entitled 'The Self of Shame':

> [...] is there a common trait shared by all shame evaluations from those that occur in other negative self-reflexive emotions? This we shall call the 'unity problem' for shame. [...] Now, some progress is perhaps accomplished if we say that, in shame, a given trait or piece of behavior is evaluated as *degrading*, or a showing one's *lack of worth*, or as having to do with an *unwanted identity*. [Deonna, Teroni 2009, p. 35]

In this last sense, *Homeland*'s case seems trapped in the identity circle more than in family relations or politics. The system of multimedia mirrors we observe appears to paralyse the subject's actions and creation of meaning. Brody sees the Brody who said it all, but this scene breaks the 'mirror rule' of identity. The dead person created in video is not him – this confession does not shape his subjectivity, because the terrorist activity did not occur. We have a situation opposite to J. M. Coetzee's statement:

> The possibility we face is of a confession made via a process of relentless self-unmasking which might yet be not [sic!] the truth but a self-serving fiction, because the unexamined, unexaminable principle behind it may be not a desire for the truth but a desire to *be a particular way*. [Coetzee 1985, p. 220–221]

In contrast, in the absence of the interrogator who leaves him alone with his own recording, Brody is faced with the need to create another self on the screen. In other words, he again confronts his 'confession self'. It is worth remembering that, in the context of the interrogation, his self-integrity becomes not only an identity issue, but a criminal one as well. Accordingly, the decision on which self Brody will choose after seeing this film has a strictly practical effect. His subjectivity can actually become synonymous with guilt. Even though the rejected self will still be a kind of the remainder of the self.

The second interrogation is totally unlike the first one. The recorded confession absolutely changed the position of the self. Rather than self-creation, we shall explore the position of guilt which comes from the idea of the self confronted with its own shame (the unwanted identity becomes the guilty identity). As a matter of fact, Brody does not have to tell the truth; he only has to elaborate on the truth. His declarations now match the importance of the Afghanistan flashbacks. The CIA verifies his previous statements – the construction of this simple confession no longer depends on him. As expected, the interrogation turns into a process of tracking the integrity of the self, and not uncovering the true self.

In conclusion, I would like to discuss the lethality of confession, a topic raised by a female character, Carrie Mathison. When she speaks to Brody, she calls his confession a 'suicide tape', which is a significant variant on the standard term (confession tape). This new term is easy to understand if we recall Brody's plans to commit suicide. Naturally, the tape will thus appear to be a confession of a future suicide. This interpretation, however, is not satisfactory. While convincing Brody to confess the truth to her, Carrie tells him that she loves him and wants him to leave his family for her. After these words she laughs: 'I said it! And I am still alive.' Through the use of a cliché or a stock phrase, the combination of confession and death reappears *en passant*. Carrie proves to her interlocutor that this connection is, firstly, unnecessary and, secondly, that his existence hinges not on the fact that he confesses, but on *what* he confesses. Therefore, one could conclude that only a certain set of 'truth confessions' is fatal – one tied to the subject's public life. Their content is irrelevant, however. Carrie's confession is not of this nature, nor does it bring these consequences, even if it is crucial for the character's development. We should rather point out that the term which she uses for Brody's confession – the 'suicide tape' – amounts to a recognition that his confession is already a form of suicide.

Let us summarise by returning to the main issue of our interpretation. The death (in this case, the suicide) of the subject occurs through his confession and the act of 'sharing' it with others. The subject gives himself up to 'make room' for the meaning he creates. After making this video, the subject no longer has the power even to oppose what he has recorded as his unmasked self (which in fact is fake or temporary). Alternately, we might say that this videotaped confession supersedes him. After all, we can ask: which self remains after this event? And how can the self be salvaged? The subject walks towards death – not only because he loses the power to construct another confession, but primarily because his actions and ability to oppose death have been limited. Lack of self-integrity is not the cause, but the effect of facing multiple 'unwanted identities' and absence of the identity one desires. This implies that the redeeming power of the confession presented by Carrie cannot apply to the present version of Brody's shattered self. His confession was to be the finale, and ultimately, it is – the death of the character occurs a few episodes later. In containing a suicide plan, the videotaped confession becomes the postponed death of the multiple uncovered selves.

Bibliography

1. Belting, H., 2014. *Faces. Historia twarzy,* trans. T. Zatorski, Wydawnictwo Słowo/Obraz Terytoria, Gdańsk.
2. Coetzee, J. M., 'Confession and Double Thoughts: Tolstoy, Rousseau, Dostoyevsky', *Comparative Literature*, Vol. 37, No. 3 (Summer, 1985), pp. 193–232.
3. Deonna, J. A., Teroni, F., 2009. 'The Self of Shame', in: *Emotions, Ethics and Authenticity*, eds. M. Salmela, V. Mayer, John Benjamins Publishing Company, Amsterdam – Philadelphia, pp. 33–50.
4. Feuer, J., 2011. 'Defining Quality Aesthetics, Form, Content HBO and the Concept', eds. J. McCabe, K. Akass, *Quality TV. Contemporary American Television and Beyond,* New York, kindle edition.
5. Gunn, J. V., 1982. *Autobiography. Toward a Poetcis of Experience,* University of Pensylvania Press, Philadelphia.

6. MacCannell, D., 1976. *The Tourist: A New Theory of the Leisure Class,* Schocken Books Inc.
7. Martin, B., 2013. *Difficult Men. Behind the Scenes of a Creative Revolution: From 'The Sopranos' and 'The Wire',* The Penguin Press, New York.
8. Thompson, R. J., 2011. 'Preface'; eds. J. McCabe, K. Akass, *Quality TV. Contemporary American Television and Beyond,* New York, kindle edition.
9. Trilling, L., 1972. *Sincerity and Authenticity,* Harvard University Press, Cambridge, Massachusetts.
10. Warchala, M., 2006. *Idea autentyczności od Rousseau do Freuda,* Wydawnictwo Universitas, Krakow.
11. Young, J. E. 'Holokaust w świadectwach filmowych i w świadectwach wideo', trans. T. Łysak, *Literatura na świecie'* 2004, No. 1–2, pp. 145–165.

Magda Ciereszko

#face, #selfie, #belfie – A new chapter in the history of the human body? Instagram as a body-centric product of *the infinity of lists*.

The article presents Instagram as a body-centric social medium with its own poetics, and one which reveals interesting tendencies. This app, launched in 2010, unquestionably changed ways of self-presentation. By sharing and hashtagging their content, Instagram users participate in an infinity of lists. Many Instagram's hashtags pertain to the body (#body, #fit, #naked) or to parts thereof (#face, #legs, #hair, etc.). However, there are also hashtags which allow people to express their physical differences (e.g. #getyourbellyout) or to join many body-oriented ideological movements, such as #freethenipple. Furthermore, Instagram's body-centric phenomenon is explored by influential artists and media. It can be seen as an example of the ambivalent state of body in modern mediated culture.

Keywords: body, Instagram, social media, selfie

Introduction: a contemporary history of the human body

The monumental three-volume synthesis *Historie du corps* [Courtine 2014] ends with a chapter about the body in the visual arts.[1] The newest history of the human body seems to be taking place on the Internet, however, and especially the Web 2.0. Social media has become an important part of people's lives. Nowadays, for many people all around the world, sharing self-generated content is part of an everyday routine, an everyday experience, and an everyday expression of themselves. It is also, or even primarily, an expression of their bodies. *Historie du corps* ends by stating that our body has become more important than our soul – it has become more important than our lives [Courtine 2014].

1 In this chapter ('Visualisations: The body and the visual arts') Yves Michaud depicts relationships between the body and visual arts in the twentieth century. Images of the body are placed in three main categories: the mechanised body, the deformed body and the beautiful body. During the twentieth century the body gained importance in art, turning into a raging obsession [Courtine 2014].

Instagram and the democratisation of photography

The popularity of social media is closely tied to another phenomenon – the democratisation of photography.[2] Nowadays, a smartphone with a camera is the standard tool. Smartphones are used for accessing the Internet and taking photos. Never before has taking photos been so simple, fast, economical and commonplace [Dziewit 2014]. We can take photos, touch them up with a special app, and share them via social media. There is no limit to these activities, we can produce and share as many photos as we want every day. As such, it is not surprising that the most popular social media platforms are mainly based on visual content.[3] They are full of images, especially amateur ones, taken on mobile phones.

One of the most important points in the history of the democratisation of photography is the invention of digital camera (1975) and the spread of this kind of photography in the 1990s. However, according to some history of photography scholars, the spread of smartphones has been even more influential than the advent of digital photography. According to Jakub Dziewit, the widespread use of smartphones is the biggest revolution not only in the history of photography, but also in the history of the human culture [Dziewit 2014]. Contemporary people spend a great deal of time on their smartphones. This device enables them to take photos and videos and share them via social media. They can also look at pictures and videos made by their friends, and even celebrities, as well as by other people they have never met. The power of images has never been so intense.[4]

---

2 The democratisation process of photography began in 1888 when the *Kodak Box* camera was invented. This model was the first camera that was easy to use and available at an affordable price. *You press the button, we do the rest* was Kodak's claim. The next revolutionary steps in the democratisation process were the premiere of the *Leica* – the first miniature camera (1925) and the invention of the digital camera (1975, though it only took off in the 1990s). Then came the market premiere of the mobile phone with a camera (2000). Nowadays, with the widespread use of smartphones, people can take photographs whenever they want, and always have a camera with them [Dziewit 2014; von Brauchitsch 2004].

3 Facebook posts are more engaging when they contain visual content [Deryło 2012]. Instagram and Snapchat only allow users to share photographs and short films. Many other social media apps have been invented to share visual content (e.g. Tumblr, Vine, Pinterest and many more).

4 Some observations from this article were developed in my other publications about social media:
Ciereszko, M. 2016. 'Performing for the followers – social media as a platform for the artist', *Contemporary Lynx*, No. 2, pp. 68–71; Ciereszko, M. 2016. 'Zabawa w "udawanego" – prowokacje artystyczne w social media', *Rynek, społeczeństwo, kultura*, No. 4, pp. 5–12;

Instagram, reality and faces

Our immersion in images is probably most perceptible on Instagram. This is a social media platform connected to a mobile app which enables its users to take and share pictures and videos. People often use it to produce images of their own. Launched in 2010, Instagram has unquestionably changed how we present ourselves. It is a real face book. Mark Zuckerberg's Facebook bought Instagram in 2012. Therefore, these two most popular social media platforms have a relationship that is both symbolic and economic.

According to the latest survey, there are over 400 million Instagram users [Woollaston 2015]. Every day, seventy million images are posted here, which means 810 photos are posted per second (Instagram 2015). Instagram has already become more popular than Twitter. The number of Instagram users in Poland exceeds three million [Dederko 2015].

Instagram has simple functions that have proved themselves to be very useful, such as image manipulation tools, location tagging of photos, and instant sharing. In a few seconds we can take a photo, modify it with filters and publish it. In the history of photography all or some of these activities have been time-consuming and laborious. This makes Instagram a labour-saving triumph. What is more, Instagram filters have become a symbol of how we project images of ourselves in our culture. They allow us to modify and correct the photos we take in a very simple way. Using filters has become so popular that Instagram users use the #nofilter hashtag when they want to announce that no filter has been used. But this facet of Instagram raises controversy. The mainstream media is full of articles on the clash between the Instagram world and reality [Sowa 2015; Tracewicz 2015]. Instagram is seen as a tool of manipulation and criticised because it creates 'ideal' images of life [Bevan 2012; Niequist 2013]. In September 2015 a photo project which exposed Instagram tricks went viral. Photographer Champoo Baritone from Bangkok posted a series of photographs on her Facebook page that showed exactly what had been cropped out of Instagram pictures [Brown 2015]. Her photographs suggest that Instagram users portray situations that have been carefully arranged, and that there is no spontaneity in photos shared via Instagram. Moreover, the tabloid online media often publishes *Instagram vs. reality* articles, in which they compare celebrities' appearances on Instagram and those captured by photographers [*Daily Mail* 2015; *Zeberka* 2015].

Ciereszko, M. 2017 (in progress) *Snapchat jako anestetyczna odpowiedź na trendy popkultury i popularyzację prywatności.*

The relationship between Instagram and reality noted by mainstream media might also be considered from a philosophical perspective. Photographs posted on Instagram can be seen as Baudrillard's *simulacra* [Baudrillard 1981]. As a tool of self-creation, Instagram divorces a photograph from reality even more than analogue and digital pictures taken with a camera. Therefore, Instagram content can be seen as a kind of hyper-reality.

Instagram is probably the most organised social media platform. Until August 2015, all the photos had the same square format and resolution (612 x 612 pixels). Now they can also be horizontal [Forbes 2015]. Videos cannot be longer than fifteen seconds and all content can be precisely marked with hashtags. A large proportion of photographs use Instagram filters, giving them a largely recognisable and standardised appearance. As a result, Instagram is a social media with its own poetic.

Hashtags – Instagram's cataloguing tools

Instagram is a catalogue. According to Umberto Eco, the Internet, in general, is 'the Mother of all Lists, infinite by definition because it is in constant evolution, the World Wide Web, which is both web and labyrinth, not an ordered tree, and which of all vertigos promises us the most mystical, almost totally virtual one, and really offers us a catalogue of information that makes us feel wealthy and omnipotent, the only snag being that we don't know which of its elements refers to data from the real world and which does not, no longer with any distinction between truth and error' [Eco 2009]. This quote comes from Umberto Eco's *The Infinity of Lists*; the title of this publication suits Instagram very well.

Instagram, just like the World Wide Web, is a special kind of catalogue which is utterly infinite, incomplete and open. Each image can be hashtagged. This simple graphic symbol (#) can be added to every word or group of words.[5] It has the power to group many images from different users. Hashtags can also be used to search for images. They allow Instagram users to find interesting and inspiring photos in a very simple way. Every hashtag forms a separate catalogue

5 The idea of using the '#' symbol before words started in 1988 on the IRC platform. At that time the name *hashtag* did not exist, but the '#' sign was used to group information, photos and videos into categories. In 2007 blogger Stowe Boyd started using #sandiegofire to group online information about fires in San Diego. He has also invented the word *hashtag*. In 2009 Twitter entered *hashtags* officially. Every word used on Twitter and preceded by a # sign become a hyperlink to the relevant category. When Instagram started in 2010, hashtags were already well known. [Troch 2014]

(e.g. #longhair), which is expanded every time it appears under a new photograph. Hashtags are also helpful in increasing the number of followers – by using catchy hashtags we can acquire followers from all over the world. Undoubtedly, Instagram users participate in an infinity of lists.

Every Instagram user can create a new hashtag, but some hashtags are used more often than others.[6] According to a report by *F5 Magazine* (F5 2014), the most popular Instagram hashtags in 2014 were:

#love – almost 144 million pictures
#instagood – 97 million pictures
#me – 80 million pictures

The third most popular hashtag, #me, could confirm the thesis that this social media platform is essentially narcissistic. Many mainstream journalists portray Instagram as a kingdom of narcissism [Chamorro-Premuzic 2014; Tylikowska 2014; Sowa 2015; Tracewicz 2015]. As a matter of fact, #me indicates its body-centric character. Many users use Instagram to produce images of their own body. As a result, the human body is a major source of content on this social media platform. Many hashtags refer to the human body (#body, #fit) or parts thereof (#face, #legs, #hair, etc.).

The infinity of lists – where the social media policy is the limit

Interestingly enough, in order to protect users against obscenity and pornography, some body-related Instagram hashtags are banned, such as #naked, #nude, #strip, #boobs, #nipples, #pussy, #penis and many more. An Instagram account can be blocked if a user posts images which do not comply by the rules. All the rules are explained in the statute which is available to users online [Instagram 2015]. Although Instagram's catalogue is restricted by the regulations, there is still space to post sexually provocative images – as in all mainstream culture.

In the context of Instagram obscenity, the history of #curvy is a very interesting case [Cliff 2015]. This hashtag was used by users to tag photos which show

6 Creating a new and unique hashtag is very important in the commercial usage of Instagram. Brands with Instagram profiles use them in online competitions. For example, Marc Jacobs created #castmemark to choose a new face for the autumn campaign of this brand [Weisman 2015]. People participating in this contest used #castmemark and their photos landed in Marc Jacob's 'catalogue'. This hashtag was also used on Twitter.
Hashtags created for commercial campaigns need to be unique and original. Hashtags like #hair, #eyes, #body are inappropriate for marketing campaigns because they are used in many other contexts [Butters 2015; Ślipko 2015].

a female body (including nude and pornographic photos). Instagram banned it in order to protect users against obscenity. But #curvy was also used by women who were plus-sized to express themselves and affirm a body which was larger than canonical images. After this hashtag was banned, XXL women fought for it, and eventually created a new one: #curvee. They criticised Instagram for its decision, noting that hashtags like #anorexia or #skinny were still allowed. As a result, Instagram unblocked #curvy and created an algorithm to check content with the hashtag for nudity or pornography.

The case of #curvy shows that Instagram is a place of symbolic disputes about the human body. The official Instagram statute includes a separate chapter on eating disorders written with the support of the National Eating Disorder Association [Instagram 2015]. This explains how to recognise an eating disorder, how to help people with this problem, and includes contact details to the relevant institutions. There is also a point about breastfeeding. Instagram allows users to post photographs depicting this activity: *We agree that breastfeeding is natural and beautiful and we're glad to know that it's important for mothers to share their experiences with others on Instagram. The majority of these photos is consistent with our rules* [Instagram 2015]. Photographs of breastfeeding, together with postmastectomy images, are evaluated positively. Instagram's statute explains this issue: *We agree that undergoing a mastectomy is a life-changing experience and that sharing photos can help awareness about breast cancer and support the women and men facing a diagnosis, undergoing treatment or living with the scars of cancer. The vast majority of these kind of photos are following our policies* [Instagram 2015]. #mastectomy has 11,395 results on Instagram.[7] Instagram's rules about eating disorders, breastfeeding and mastectomy correspond with Foucault's concept of the body in the power of discourse [Foucault 1975; 1976].

Instagram's statute also contains a chapter on sexting. It warns users against risky online behaviour and provides advice on what to do when somebody requests nude photographs or videos. This chapter was developed in cooperation with ConnectSafely.org, a global organisation whose main focus is to prevent sexting. In this section we find the statement that *sharing nude or sexual photos/ videos is a violation of Instagram's Community Guidelines* [Instagram 2015].

Nevertheless, many Instagram users share their erotic photos. Profiles with this kind of content are very popular. For example, according to the *Napoleon*

7 The number of followers for 22.10.2015. #mastectomytatoo – 1,394 results, #mastectomybras – 279 results [Instagram 2015].

Cat survey,[8] the second most popular Pole on Instagram (the first is Robert Lewandowski with 3.2 million followers[9]) is Joanna Kuchta (695,000 followers),[10] a nineteen-year-old model from Poland living in Ireland.[11] She posts many photos of herself which are erotically charged and provocative, especially in her use of the Lolita image. Joanna exhibits her feminine body in girl's underwear and poses in an erotic fashion. She mixes cute juvenile colours, childish gestures and gadgets (such as hair clips, bows and rabbit ears) with signifiers of the 'adult world', such as collars, fishnets, lipstick, and stockings. Some of her photographs have been blocked by Instagram, but she remains active and popular in this social media.

When it comes to the world's most popular profiles on Instagram, second place belongs to Kim Kardashian West, with 49.7 mln followers[12] [Instagram 2015]. This American celebrity is well known for posting photos that exhibit her buttocks. Despite posting this kind of content, she has never been blocked by Instagram, while the profiles of supermodel Anja Rubik and pop star Rihanna were blocked for some time when they exposed their nipples [Paul 2014; Syc 2014]. These examples reveal the inconsistencies of Instagram's policies. They also correspond with the discourse of body control: condoned and prohibited forms of bodily expression are significant in culture [Banaszak 2012]. Of course, the rule prohibiting photographs that show female nipples is a result of Instagram's American provenience. This app was created in the USA, and in American culture female nipples are especially obscene [Yalom 2012]. There is no denying that Instagram's policies derive from North America's cultural approach to the body.

---

8 *Napoleon Cat* is a Polish start-up specialising in social media analysis. It is a tool for people managing commercial social media profiles, helping them in their publishing, analysing and optimising of social media. It delivers and analyses data from Facebook, Google Plus, Twitter, Instagram and YouTube. To promote its tool, *Napoleon Cat* regularly ranks the most popular profiles of brands, institutions and people on social media. The list cited in the article was published in June 2015 [Berezowski 2015].

9 The number of followers on 22.10.2015 [Instagram 2015].

10 The number of followers on 22.10.2015 [Instagram 2015].

11 Joanna is a face of *Anti-agency* – a modeling agency which states that *the agency is for people who could've been models and decided not to, for people who are too cool to be models and people with real lives on the verge of exploding in music, fashion, art, illustration & creative industries etc* [Anti-agency 2015]. The faces of *Anti-agency* very often collaborate with streetwear brands (eg. *American Apparel*).

12 The most popular celebrity on Instagram is Taylor Swift, with fifty-two million followers. For quite a long time first place belonged to Kim Kardashian West; the change in this ranking was analysed by mainstream media [Sinha-Roy 2015].

As previously mentioned, Instagram allows users to post photos of breastfeeding and postmastectomy. However, other images of female breasts with bare nipples are banned. In response to this policy, the #freethenipple campaign was born.[13] It started on Twitter, but it has also gained ground on Instagram. It is a movement against double standards of human body exposure in social media: while men can show their nipples, female breasts are restricted by regulations. This is, of course, a transfer of 'offline' ideas – since the 1960s women have been exposing breasts in a political context and fighting for the right to appear topless in public [Yalom 2012, pp. 291–328]. In *A History of the Breast,* the chapter about those efforts concludes with Marilyn Yalom asking whether women will ever acquire the right to bare their breasts in public [Yalom 2012]. It is certain that social media, especially Instagram, is a very important venue for achieving this end.[14]

Another important case demonstrating the discursive context of the body in Instagram is the story of the photo of Rupi Kaur – a young Canadian artist and poet. The photograph depicts Rupi reclining with traces of menstrual blood on her trousers and bed sheet. It was removed by Instagram's administrators twice. This image was part of her *Period* project prepared for a university visual rhetoric course. To protest, Rupi Kaur reported this incident on her Facebook and Tumblr accounts. She accused Instagram of misogyny and of favouring the patriarchial order. The posts were powerful – they went viral, and as a result Instagram apologised to Rupi Kaur and reinstated the controversial photo [Gray 2015]. This case exposes the enduring menstruation taboo in mainstream culture – the image of

---

13 Women supporting #freethenipple campaign (started by the activist and filmmaker Lisa Esco) oppose double standards that lead to the censorship of female breast [Esco 2014; Ferrier 2015]. On Facebook and Instagram sharing a graphic content with a woman's aerola is prohibited while male nipples are not. To undermine these double standards many Instagram users joined the #freethenipple campaign and posted photographs with exhibited nipples using the hashtag. Some of them were very creative. For example, artist Micol Hebron posted a series of photos on Instagram in which her nipples are obscured by images of men's nipples cropped from the other photos. The *Free the nipple* campaign has its own official Instagram profile with 225,000 followers [Instagram, 2015]. The campaign is supported by many celebrities, e.g. Miley Cyrus, Rihanna, Anja Rubik.

14 I wrote about Free the nipple as an example of countercultural movement in social media in: Ciereszko, M., 2016. 'Performing for the followers – social media as a platform for the artist', *Contemporary Lynx*, No. 2, pp .68–71.Ciereszko, M. 2016. '„Zabawa w udawanego" – prowokacje artystyczne w social media', *Rynek, społeczeństwo, kultura*, No. 4, pp. 5–12.

menstrual blood is prohibited even in tampon and maxi pad commercials. Social media policies that uphold this taboo are among its contemporary symptoms.[15]

Rupi Kaur's and Lina Esco's ideas are live on in social media – Twitter has a popular #liveyourperiod campaign, and on Instagram the #freethenipple campaign continues unabated. People use these hashtags to express their views of the body in the contemporary culture.

Instagram as a body catalogue

Social media emanations of the body are important for building users' identities in many ways. Using Instagram hashtags seems to be a way of expressing affiliation with a certain community. Hashtags allow people to place their bodies in desirable groups (e.g. #beauty, #pretty, #young, #fitness). This is a simple way of showing off.[16] Alise Tifentale summed up this mechanism as follows: *The individual and unique #me becomes part of #us, a virtual community via means of a common platform for image sharing and the uniform image format provided by Instagram* [Tiefentale 2013]. On Instagram the postmodern concept of the body as a *text* to be given significance [Bauman 1995] acquires literal meaning. As mentioned above, this app is a body-centric medium. Hashtags pertaining to the body make Instagram an incomplete, infinite and open catalogue of images of the human

15 I analysed the case of Rupi Kaur in a context of art in social media in : Ibidem.

16 *Showing off* is a type of behavior which is often mentioned in the context of social media. It can be defined as boasting of material goods and pastimes. It is ostentatious consumption. According to the qualitative analysis prepared for Polish marketing blogger Natalia Hatalska, *showing off* takes place more often on Instagram than it does on Facebook [Hatalska 2014]. This results from the fact that Instagram's functions are based on sharing photos. The only available form of visual (or audiovisual – a short video option is also available) presentation encourages users to share information about material goods. Furthermore, an Instagram account can be more anonymous than a Facebook account (users are not obliged to use their real name and are not encouraged to provide many details about themselves), which minimises users' tendency to censor themselves. Instagram users give less thought to their friends' opinions about their content sharing than Facebook users do. As a result, the Instagram is a social media platform for presenting users' 'material' face (their appearance, the goods they own), while Facebook is a tool for expressing their inner selves – their interests, passions, or special activities [Hatalska 2014]. A survey prepared by Natalia Hatalska discovers interesting differences between the creation of self-images on Instagram and on Facebook. It must be noted, however, that this distinction is subtle and correlates with the type of content (Instagram permits only photographs, graphics or videos) which can be shared in these social media.

body. This collection has an interesting structure, logic and dynamics which deserve scholarly attention.

Instagram as a body catalogue: parts of the body

On Instagram the body is often fragmented. There are hashtags in common use that pertain to the parts of the human body, such as #face, #hair, #eyes, #lips, #legs, #stomach, etc. In these Instagram body images, people exhibit parts of their body for various reasons. Sometimes they present something unusual connected to that part of their body (for example, somebody has hurt his finger and wants sympathy); at other times this is a way of showing off (for example, when a girl wants to display her beautiful hair after a visit to the hairdresser). This is particularly true of athletic people, such as fitness trainers, who display their bodies on Instagram in order to motivate others. They use hashtags that pertain to well-honed parts of their bodies, such as #abs or just #muscles. This kind of 'body hashtag' is often used by people searching for 'body inspirations'. Women and men look for interesting haircuts using #hair, people who need motivation for work-outs use #abs, #muscles, etc. This facet of Instagram users' behaviour is used in commercial and social marketing campaigns. The senders of marketing content often use popular body hashtags to locate their product or idea in a desirable context [Ślipko 2015]. This can be as trivial as using #hair to situate a photo of a new hairspray in the results of #hair, but it can be eye-catching and creative when used in a charity campaign, for example.[17]

Instagram as a body catalogue: types of the body

In posting pictures of their body, Instagram users can also use more general hashtags, allowing them to express their belonging to a category. Hashtags such us #fitness, #beauty and #pretty assign the user's body to a desirable group, reflecting their aspirations. But there is also a group of hashtags used to affirm body images which are taboo or controversial in contemporary culture. They help users accept their bodies and give them a feeling of belonging to a group of like-minded people.

#BodyHonest is a hashtag used to show a natural body with spots, scars, freckles, etc. People use it as a response to the popular images of the ideal body posted

17 As mentioned above, the hashtag of a marketing campaign should be unique and original, but sometimes marketers only need a popular hashtag's support to increase the visibility of their content. As such, using a popular hashtag like #hair is only an auxiliary tool [Ślipko 2015].

on Instagram by celebrities, models, bloggers, fitness trainers and ordinary people with 'ideal beauty' inspirations. There are many more popular hashtags 'against' ideal body images promoted in Western mass media such as [Dupere 2015]:

- #thighreading (photos exposing thigh imperfections, such as cellulite and stretch marks)
- #loveyourlines (stretch marks)
- #plusize, #curve, #effyourbeautystandards (photos of women who are bigger than 'standard'), #fatkini (photos of plus-size women wearing bikinis)
- #BlackIsBeautiful (a black empowerment slogan since the 1960s) and #Melanin

All of the above-mentioned hashtags question existing discourses of the body, such as the discourse of weight-loss [Melosik 2010]. Furthermore, for minorities such as the disabled, transgender or people with serious diseases, using Instagram may be a path to accepting their bodies. Whether due to size or disability, physical appearance can be a major source of discrimination. A sense of community matters here, too. For example, wheelchair users have created the hashtag #wheelchairlife to help in the fight against the stigma of disability. They show happy people with mobility-related disabilities. This hashtag debunks the common myth that people living with disabilities are somehow inferior to able-bodied people, which is a message many still need to hear. Another affirmative hashtag of the 'body' minority is #TransIsBeautiful. This allows transgender people to stress their attractiveness, which might be less than obvious to the rest of society, and to fit into the beauty discourse.

The body on a life support system is still a taboo image in a mainstream culture. The popular media avoids showing people with stoma unless it is a charity campaign which aims to shock the audience. But for people suffering from Crohn's disease, a stoma is part of their everyday life, and part of their everyday struggle with the body. To destigmatise the use of colostomy bags, #GetYourBellyOut encourages social media users to show off this medical apparatus as a sign of strength and pride. A similar gesture has been made for diabetes using insulin pumps. #ShowMeYourPump encourages people to show their health equipment and to appreciate it [Dupere 2015].

#wheelchairlife, #TransIsBeautuful, #GetYourBellyOut and #ShowMeYourPump are interesting examples of building a community around body status on Instagram. These examples of hashtags fight the stereotype of Instagram as a contemporary medium that supports the demanding Western beauty canon and a space to show off. These hashtags expose material items and improve the body with the app's filters. As it turns out, this social media platform also allows its users to manifest their physical dissimilarity. It is a space to express the individual identities of users' bodies and to question the contemporary canons of beauty. Sexual minorities, people with disabilities, people affected by various diseases and physical defects can find a sense of community, make friends and enhance their self-esteem on Instagram. Presenting this social media platform exclusively in the context of the cultural narcissism which is a common praxis of the mainstream media is partial and conventional. Promoting acceptance of yourself and of your body is also part of Instagram.

Instagram as a body catalogue: *selfies*

Without a doubt, the most recognisable kind of depiction of the human body associated with Instagram is the *selfie* – a self-portrait taken with a mobile phone. The face is usually, but not necessarily in the centre of the photograph.

Since November 19, 2013 when Oxford Dictionaries announced *selfie* as 'the International Word of the Year', this hybrid phenomenon of vernacular photography and social media has created quite a bit of media hype [CNN 2013]. According to Jenna Wortham, technology reporter for the *New York Times*: 'Selfies have become the catchall term for digital self-portraits abetted by the explosion of mobile phone cameras, photo-editing and sharing services. Every major social media site is overflowing with millions of them. Everyone from the pope to the Obama girls has been spotted in one' [Wortham 2013].

The word *selfie* is defined by the Oxford Dictionary as: 'a photograph that one has taken of oneself, typically one taken with a smartphone or webcam and shared via a social media website' [Oxford Dictionaries 2013]. Therefore, the selfie is connected with social media, and the genesis of this type of photo can take its place in the history of the self-portrait. The person who is often cited as having taken the first photographic self-portrait (1839) is Robert Cornelius [Żelazińska 2014]. However, heavy and cumbersome cameras did not favour self-portraits. Lightweight and easy-to-use digital cameras and Polaroids popularised this type of photography. Nevertheless, the most important development for this contemporary phenomenon is the widespread use of smartphones [Dziewit 2014]. It might be said that the selfie is the most massive form of self-reproduction, unprecedented in history [Poprzęcka 2013]. Never before have so many people owned a camera – a small device that is always at the ready.

An interesting chapter in the history of cameras built into smartphones is the fact that the camera was initially located at the back of the smartphone. This meant that people who tried to take self-portraits had to experiment – they could not see themselves while taking this kind of photograph. Then a little mirror was installed at the back of the smartphone, but many people thought it was merely a piece of decor and did not use it. Nowadays, smartphones have two cameras – one at the back and one at the front. This makes taking self-portraits much easier. It turns out that people love these kinds of photographs, and that they need this mode of self-expression [Dziewit 2014, pp. 190–192].

As Alice Tifentale noted in an article [Tifentale 2014], 'selfies have been called "a symptom of social media-driven narcissism" [Pearlman 2013], a "way to control others' images of us," a "new way not only of representing ourselves to others, but of communicating with one another through images" [Rawlings 2013], "the masturbation of self-image" [Marche 2013] and a virtual "mini-me," that in ancient biology might have been called a "homunculus" – a tiny pre-formed person that would grow into the big self'. Łukasz Dziewit sees a selfie as a way of participating in everyday reality and points out an interesting difference between an 'ordinary' self-portrait and a selfie [Dziewit 2014]:

> In the case of the self-portrait the "self" is the most important, it tells everyone the **aim** of this photo. Meanwhile, in a selfie the person is less important than his or her designation: this is me **somewhere**, it is me **looking like this**, that is me **with someone**. It is possible that photographs have become necessary in order to look any way at all, be with anyone or be anywhere.[18]

Therefore, a selfie is a type of photograph rooted in everyday experiences, which requires neither effort nor preparations. Even if someone looks worse than usual, a selfie can be corrected by filters. Furthermore, it is easy to find Internet guides on how to make the perfect Instagram selfie. The queen of this kind of self-portrait is considered to be Kim Kardashian and her sisters. Kim Kardashian's book of her best selfies was released in May. The popularity of the selfie might seem baffling, but Tomasz Szlendak's explanation sounds convincing: 'Possibility is the stimulus for taking a selfie. Until recently, taking pictures and publishing photos was a privilege of the few. Nowadays, everyone can do it. And so they do. Without thinking' [Sowa 2015].

The selfie is a hybrid that requires hybrid methodological approaches. *Selfiecity* is a research project run by Dr Lev Manovich and the Software Studies Initiative [Selfiecity 2014]. It is an attempt to make sense of the multitude of selfies posted on Instagram. This project investigates the styles of self-portraits (selfies) in five cities across the world (Bangkok, Sao Paolo, Berlin, Moscow, and New York). The team led by Lev Manovich presented their opinions on the demographics of people

18 This definition of selfie and the results of the Selficity project were used in my article: Ciereszko, M. 2017 (in progress) 'Snapchat jako anestetyczna odpowiedź na trendy popkultury i popularyzację prywatności'.

taking selfies. The interactive 'selfiexploratory' allows us to navigate the whole set of 3,200 selected photos. In *Selfiecity*, the selfie is treated as individual Instagram users' self-expression, as well as a communal and social practice. The research project considers the individual artistic intentions of a single image, as well as the overall patterns revealed by the large amount of selfies taken in a particular geographic location during one week.

The team downloaded Instagram photographs that were taken during one week in October 2013 and geo-tagged them in the central areas of five cities: Bangkok, Berlin, Moscow, New York, and Sao Paulo. 140,000 photographs were then randomly selected for further analysis. Three Mechanical Turk workers were assigned to identify the selfies from this pool of images. Their work, double-checked by the team members, resulted in the selection of 640 selfies from each city.

The basic question of this survey was: What can a group of selfies taken in a particular city possibly tell us about it? The members of the survey team analysed the age of the selfie creators, the composition of these images and the moods of the people being photographed. Interestingly enough, according to the *Selfiecity* survey, people take fewer selfies than is often assumed. Depending on the city, only 3–5% of analysed images were actually selfies. This does not correspond with the popular thesis of the narcissistic nature of social media. The masturbation in self-portraits is a view that can be seen even in the vulgar Polish word for selfie, which is 'samojebka' – a combination of the words for 'self' and 'fuck'.

Instagram as a body catalogue: 'post' selfies

Selfies have unquestionably become a part of popular culture. The gesture of taking a selfie (a raised hand holding a smartphone or selfie stick) is probably as recognisable as that of taking a photo with a camera (pressing the button which snaps). So the time has come for 'post-selfies'. Instagram contains many selfie hybrids, such as the #belfie (focusing on the buttocks) or the #bumpie (exposing a pregnant woman's belly). The tendency to expose parts of the body in tight close-ups corresponds with Brian McNair's conception of the striptease culture [McNair 2004]. Furthermore, postmodern *impression collectors* [Bauman 1995] post extreme selfies on Instagram, taken on the roofs of buildings, on the tops of rocks, and in dangerous circumstances. From time to time the media reports the deaths of people who attempted to take extreme selfies [Rizzo 2015].

Instagram as a body catalogue: seasonal body obsessions

Instagram users also create some odd body trends. There are many hashtags for specific kinds of body representations which were born in this social media platform. One of them is #thighgap. This is the space between the inner thighs of women when standing with the knees touching. This hashtag was born in 2012 after a Victoria's Secret fashion show. Pictures tagged with #tighgap appeared in the social media. The hashtag and the image have become so popular that even nutritionists and doctors have commented on this social media trend [Siassina 2013; Greenberg 2013]. In their opinion, fixation on thigh gap can lead to eating disorders and self-esteem issues, because not every woman is able to conform her body to this trend.

Another hashtag, #bikini bridge, denotes the gap which appears between the pants and the body when a woman with a flat stomach lies down and her hip bones protrude. This hashtag was extremely popular in 2014 [Daily Mail 2014]. The newest Instagram buzzword is thighbrow [Song 2015]. This pertains to a set of folds that frames the tops of the thighs and separates the legs from buttocks. As with #thighgap, the bikini bridge and the thighbrow images have been widely discussed and criticised. Feminists have noted that these kinds of social media images can be oppressive for women [Mansour 2013; Swash 2013]. The media has also criticised the Belly Button Challenge (#bellybuttonchallenge) – many people posted photos in which they try to stretch an arm behind their back and around to touch their belly button [Knoblauch 2015]. A successful attempt means that somebody is 'slim enough'.

Instagram and the postmodern body

Instagram images of the human body also support the postmodern conception of the body as a project [Bauman 1995]. Through filters, Instagram users can manipulate pictures and change their images, creating various versions of themselves. To give their body images the label they desire, Instagram users can use the appropriate hashtags. The category of *fitness*, which is key to the postmodern theory of the body, is highly visible on Instagram, and not only in challenges like the Belly Button Challenge. Many Instagram users posting their photos seek to prove that their bodies are *fit,* which means: 'sensitive, absorbent, tuned to the absorption of pleasure, all sorts of pleasures – sexual, gastronomic, visual and auditory, but primarily – pleasure of having the body which is trained in the art experience pleasure. 'Fitness' is not about the body's ability to feats, it's about the depth of impressions during them. It's about impressions – they should be "exiting", "delightful", "charming", "ecstatic". [Bauman 1995, pp. 91–92].

Instagram's body representations seem to be very expressive. So, is there an Instagram canon of beauty? Certainly, new image-making and image-sharing technologies demand radically new ways of looking at the images. Instagram has created a few recognisable ways of presenting the body, such as the selfie, belfie, bikini bridge and thigh gap. We may also note that Instagram stars very often pose in a characteristic fashion: the face is in the centre, the cheeks are sucked in, the eyes are bulging and the lips are puckered. Instagram even set a trend for a new kind of make-up that creates very large lips (this effect is achieved using a lip liner that expands the natural contours of the lips). However, it is probably too early to draw any general conclusions. Because of its enormous scale, Instagram body images are very diverse, and so we must be careful with generalisations.

Instagram – a new chapter in the history of the human body?

Paraphrasing the question of *Selfiecity* project: does Instagram disclose anything new about our culture? The answer may be disappointing: no, it does not. The desire to be seen and the need for self-presentation are merely parts of human nature. Even the selfie can be considered a continuation of the tradition of the self-portrait in painting and photography.

Nevertheless, Instagram is a body-centric social medium with its own poetics, and one which reveals interesting tendencies. The fragmentation of the human body and centralisation of the face occur at the same time. Users participating in Instagram's *infinity of lists* not only show off their bodies, as is customary in Western culture, but they also express their physical differences or join many body-oriented ideological movements such as #freethenipple. This social media platform gives people a tool for bodily expression which is literally at your fingertips. This is a fact which should be noted in the contemporary history of the human body – maybe as a new chapter. On the surface, Instagram could be apprehended as an example of the ambivalent condition of the human body in a modern mediated culture.

Bibliography

Print

1. Banaszak, E. and Czajkowski P. and Florkowski R., 2012. *Fenomeny kontroli ciała*. Warszawa. Difin.
2. Banaszak, E. and Czajkowski, P. eds., 2010. *Corpus delicti, rozkoszne ciało. Szkice nie tylko z socjologii ciała*. Warszawa. Difin.
3. Banaś, M. and Warmińska K. eds., 2011. *Kulturowe emanacje ciała*, Kraków. Wydawnictwo Uniwersytetu Jagiellońskiego.
4. Baudrillard, J., 1981. *Symulakry i symulacja*. Translated by Sławomir Królak. 2005. Warszawa. Wydawnictwo Sic!
5. Bauman, Z., 1995. *Ciało i przemoc w obliczu ponowoczesności*. Toruń. Uniwersytet Mikołaja Kopernika.
6. Belting, H., 2014. *Faces: historia twarzy*. Translated by Tadeusz Zaorski. 2015. Gdańsk. Wydawnictwo Słowo/Obraz Terytoria.
7. Carr-Gomm, P., 2010. *Historia nagości*. Translated by Agnieszka Wyszogradzka-Gaik. 2010. Warszawa. Bellona.
8. Courtine, J. ed., 2006. *Historia ciała. T. 3. Różne spojrzenia: wiek XX*. Translated by Krystyna Belaid and Tomasz Stróżyński. 2014. Gdańsk. Wydawnictwo Słowo/Obraz Terytoria.
9. Dziewit, Ł., 2014. *Aparaty i obrazy. W stronę kulturowej historii fotografii*, Katowice. Grupakulturalna.pl.
10. Eco, U., 2009. *Szaleństwo katalogowania*. Translated by Tomasz Kwiecień. 2009. Poznań. Dom Wydawniczy Rebis.
11. Foucault, M., 1975. *Nadzorować i karać. Narodziny więzienia*. Translated by Tadeusz Komendant. 2009. Warszawa. Wydawnictwo Aletheia.
12. Foucault, M., 1976. *Historia seksualności*. Translated by Bogdan Banasiak, Tadeusz Komendant, Krzysztof Matuszewski. 1995. Warszawa. Czytelnik.
13. Gromkowska-Melosik, A. ed., 2007. *Kultura popularna i (re)konstrukcje tożsamości*. Poznań-Leszno. Wyższa Szkoła Humanistyczna.
14. McNair, B., 2002. *Seks, demokratyzacja pożądania i media, czyli Kultura obnażania*. Translated by Ewa Klekot. 2004. Warszawa. Muza.
15. Melosik, Z., 2010. *Tożsamość, ciało i władza w kulturze instant*. Kraków. Oficyna Wydawnicza Impuls.
16. Musiał, M., 2015. *Intymność i jej współczesne przemiany: studium z filozofii kultury*. Kraków. Towarzystwo Autorów i Wydawców Prac Naukowych Universitas.
17. Poprzęcka, M., 2014. 'Ludzkości portret własny', *Książki*, 3, pp. 8–11.

18. Sowa, A., 2015. 'Fotkowe zapalenie mózgu', *Polityka*, 19, pp. 22–25.
19. Sowa, A., 2015. 'Fotkowe zapalenie mózgu', *Polityka*, 19, pp. 22–25.
20. Szubrycht, J., 2014. 'Na wyciągnięcie ręki', *Polityka*, 12, pp. 97–99.
21. Tracewicz, J., 2015. 'Wakacje na pokaz', *Polityka*, 33, pp. 15–18
22. Tracewicz, J., 2015. 'Wakacje na pokaz'. *Polityka*, 33, pp. 15–18.
23. von Brauchitsch, B., 2002. *Mała historia fotografii*. Translated by Jan Koźbiał, Barbara Tarnas. 2004. Warszawa. Wydawnictwo Cyklady.
24. Yalom, M., 1998. *Historia kobiecych piersi*, 2012. Warszawa. Wydawnictwo ZIELONE DRZEWO Instytut Psychologii Zdrowia
25. Żelazińska, A. 2014. 'Selfie – portret', *Ja – My – Oni. Co komputer zrobił nam z głową? Poradnik Psychologiczny Polityki*, 16, pp. 28–32.

Online articles and social media

Bevan, K., 2012. 'Instagram is debasing real photography'. *The Guardian*, [online] 19 July, Available at: <http://www.theguardian.com/technology/2012/jul/19/instagram-debasing-real-photography> [Accessed 15 September 2015].

Berezowski, G., 2015. 'Ranking polskich profili na Instagramie – marzec 2015'. *NapoleonCat.com*, [online] 6 April. Available at: <https://napoleoncat.com/blog/ranking-polskich-profili-na-instagramie-marzec-2015> [Accessed 21 September 2015].

Binczycka A., 2014. 'Fenomen selfie, czyli jak wypromować samego siebie'. *Marketing przy kawie*, [online] Available at: <http://www.marketing-news.pl/theme.php?art=1765> [Accessed 20 September 2015].

Brown, S., 2015. 'Photographer Shows The Reality Behind Instagram Photos In Eye-Opening Series'. *The Huffington Post*, [online:] Available at: <http://www.huffingtonpost.co.uk/2015/09/19/chompoo-baritone-instagram_n_8162652.html> [Accessed 2 September 2015].

Brumfield, B., 2013. 'Selfie named word of the year for 2013'. *CNN*, [online] 20 November. Available at: <http://edition.cnn.com/2013/11/19/living/selfie-word-of-the-year> [Accessed 29 September 2015].

Butters K., '#HashtagMarketing: jak używać hashtagów by dotrzeć do właściwej grupy odbiorców', *Get Response*, [blog] 29 October. Available at: <http://blog.getresponse.pl/hashtagmarketing-uzywac-hashtagow-dotrzec-wlasciwej-grupy-odbiorcow.html> [Accessed 22 September 2015].

Chamorro-Premuzic, T., 2014. 'Sharing the (self) love: the rise of the selfie and digital narcissism'. *The Guardian*, [online] 13 March. Available at: <http://www.theguardian.com/media-network/media-network-blog/2014/mar/13/selfie-social-media-love-digital-narcassism> [Accessed 12 October 2015].

Cliff, M., 2015. Bringing #curvy back: Instagram lifts ban on search term after backlash from plus-size women outraged over ruling because 'skinny' and 'anorexia' were still allowed. *Daily Mail,* [online] 27 July. Available at: <http://www.dailymail.co.uk/femail/article-3175960/Bringing-curvy-Instagram-lifts-ban-search-term-backlash-plus-size-women.html> [Accessed 26 September 2015].

Daily Mail, 2014. 'Has "bikini bridge" become the new thigh gap? Disturbing new selfie fad circulating on social media'. *Daily Mail,* [online] 7 January. Available at: <http://www.dailymail.co.uk/femail/article-2535098/Is-bikini-bridge-new-thigh-gap-Disturbing-new-selfie-fad-circulating-social-media.html> [Accessed 30 September 2015].

Daily Mail, 2015. 'It's an Insta-sham! From Beyonce to Kim, the stars who look a little different in real life (even on the same day)'. *Daily Mail,* [online] 20 March. Available at: <http://www.dailymail.co.uk/femail/article-2988263/It-s-Insta-sham-Britney-Spears-Beyonce-Kim-Kardashian-Miranda-Kerr-favourite-stars-slimming-Instagram-photographs.html> [Accessed 5 September 2015].

Dederko J., 2015. 'Ponad 3 mln polskich użytkowników na Instagramie, większość to kobiety i nastolatki', *Social Press,* [online] Available at: <http://socialpress.pl/2015/09/ponad-3-mln-polskich-uzytkownikow-na-instagramie-wiekszosc-to-kobiety-i-nastolatki> [Accessed 2 October 2014].

Deryło, J. 2012, 'Image is the king, czyli krótka historia facebookowej dominacji obrazu'. *Spider's Web Blog,* [blog] 27 July. Available at: <http://www.spidersweb.pl/2012/07/image-king-czyli-krotka-historia-facebookowej-dominacji-obrazu.html> [Accessed 19 September 2015].

Dupere, K., 2015. '9 hashtags that champion body positivity for all'. *Mashable.com,* [online] 1 August. Available at: <http://mashable.com/2015/08/01/body-positivity-hashtags/#n4bOSfG4duq9> [Accessed 30 September 2015].

Esco, L., 2014. 'Facebook Wages War on the Nipple'. *The Huffinton Post,* [online] 1 July. Available at: <http://www.huffingtonpost.com/lina-esco/facebook-war-on-nipples_b_4548832.html> [Accessed 14 September 2015].

F5, 2014. 'Trendy na Instagramie: 100 najpopularniejszych hashtagów'. *F5,* [online] 15 October, Available at: <http://www.fpiec.pl/post/2014/10/15/trendynainstagramie100najpopularniejszychhashtagow> [Accessed 21 September 2015].

Ferrier, M., 2015. 'How to beat Instagram and Facebook's female nipple ban? Use male nipples instead'. *The Guardian,* [online] 7 July. Available at: <http://www.theguardian.com/lifeandstyle/shortcuts/2015/jul/07/instagram-facebook-female-nipple-ban-use-male-nipples-instead> [Accessed 19 September 2015].

Forbes.pl, 2015. 'Koniec z kwadratem. Instagram dodaje nowe formaty zdjęć', *Forbes,* [online] 28 August 2015. Available at: <http://www.forbes.pl/instagram-dodaje-nowe-formaty-zdjec-koniec-z-kwadratem,artykuly,198539,1,1.html#> [Accessed 1 September 2015].

Gray, E., 2015. 'The Removal Of Rupi Kaur's Instagram Photos Shows How Terrified We Are Of Periods'. *The Huffington Post,* [online] 27 March. Available at: <http://www.huffingtonpost.com/2015/03/27/rupi-kaur-period-instagram_n_6954898.html> [Accessed 30 September 2015].

Greenberg, B., 2013. 'The Thigh Gap – A Disturbing New Trend Among Teen Girls'. *The Huffington Post,* [online] 13 April, Available at: <http://www.huffingtonpost.com/barbara-greenberg/the-thigh-gapa-distubing-_b_2647646.html> [Accessed 30 September 2015].

Hatalska, N., 2014. 'Zjawisko show off w mediach społecznościowych: analiza wypowiedzi internautów, badanie zrealizowane przez Replise Poland na zlecenie hatalska.com'. *Natalia Hatalska Blog,* [blog] 8 January. Available at: < http://hatalska.com/2014/01/08/pokolenie-show-off> [Accessed 26 September 2015].

http://www.oxforddictionaries.com/definition/english/selfie [Accessed 29 September 2015].

Instagram's policies, 2015. *Instagram.com,* [online] Available at: <https://help.instagram.com/478745558852511> [Accessed 30 October 2015].

Knoblauch, M., 2015. 'People are reaching behind their backs to touch their belly buttons, and taking photos to prove it'. *Mashable.com,* [online] 12 June. Available at: <http://mashable.com/2015/06/12/belly-button-challenge/#RdzM.V0ddSqm> [Accessed 30 September 2015].

Mansour, I., 2013. '"Thinspiration" Packages Eating Disorders as a Lifestyle Choice'. *Mashable.com,* [online] 5 December. Available at: <http://mashable.com/2013/12/05/thinspiration/#HfJ8q4afBEqk> [Accessed 30 September 2015].

Marche, S., 2013. 'Sorry, Your Selfie Isn't Art'. *Esquire,* [online] 24 July. Available at: <http://www.esquire.com/entertainment/a23557/selfies-arent-art> [Accessed 30 October 2015].

Marczak, G., 2013. 'Dawno, dawno temu, czyli skąd masz aparat w telefonie', *AntyWeb Blog,* [blog] 19 November. Available at: <http://antyweb.pl/dawno-dawno-temu-czyli-skad-masz-aparat-w-telefonie> [Accessed 15 September 2015].

Niequist, S., 2013. 'Instagram's Envy Effect'. *Relevant Magazine,* [online] 30 August, Available at: <http://www.relevantmagazine.com/culture/tech/stop-instagramming-your-perfect-life> [Accessed 10 September 2015].

Paul, K., 2014. 'She's back: Rihanna returns to Instagram after 6-month absence'. *Mashable.com,* [online] 2 November. Available at: <http://mashable.com/2014/11/02/rihanna-back-instagram/#YDWB0c.DnSq6> [Accessed 29 September 2015[.

Pearlman, J., 2013. 'Australian man "invented the selfie" after drunken night out'. *Daily Telegraph,* [online] 19 November. Available at: <http://www.telegraph.co.uk/news/worldnews/australiaandthepacific/australia/10459115/Australian-man-invented-the-selfie-after-drunken-night-out.html> [Accessed 30 October 2015].

Rawlings, K., 2013. 'Selfies and the history of self-portrait photography'. *OUP Blog,* [blog] 21 November. Available at: <http://blog.oup.com/2013/11/selfies-history-self-portrait-photography> [Accessed 28 October 2015].

Rizzo, C., 2015. 'More people have died from selfies than shark attacks this year'. *Mashable.com,* [online] 21 August. Available at: <http://mashable.com/2015/09/21/selfie-deaths/#8EiHomwEskqM> [Accessed 29 September 2015].

Selfiecity, 2014. [online] Available at: <http://selfiecity.net/#> [Accessed 5 September 2015].

Siassina, M., 2013. '"Thigh gap" measuring trend among young women on social media may be encouraging eating disorders'. *National Post,* [online] 9 October. Available at: <http://news.nationalpost.com/health/thigh-gap-measuring-trend-among-young-women-on-social-media-may-be-encouraging-eating-disorders-u-s-experts> [Accessed 28 September 2015].

Sinha-Roy, P., 2015. 'Taylor Swift, Kim Kardashian lead most-followed Instagram accounts', *Reuters.com,* [online] 6 October. Available at: http://www.reuters.com/article/2015/10/06/us-instagram-taylorswift-idUSKCN0S012L20151006> [Accessed 15 October 2015].

Song S., 2015. 'Thigh brows are the newest celebrity Instagram Trend'. *PaperMag,* [online] 14 September. Available at: <http://www.papermag.com/2015/09/thigh_gaps_are_out_thigh_brows.php> [Accessed 15 October 2015].

Swash, R., 2013. 'How the "thigh gap" became the latest pressure point on a woman's self-image'. *The Guardian,* [online] 3 November. Available at: <http://www.theguardian.com/lifeandstyle/2013/nov/03/thigh-gap-pressure-point-women-self-esteem> [Accessed 22 Septmeber 2015].

Syc, M., 2014. 'Czy Instagram dyskryminuje puszyste osoby?' *Social Press,* [online] 17 July. Available at: <http://socialpress.pl/2014/07/czy-instagram-dyskryminuje-puszyste-osoby> [Accessed 20 September 2015].

Ślipko, S., 2015. 'Instagram – hashtagi z głową', *Media & Marketing Polska,* [online] 23 March. Available at: <http://mediamarketingpolska.pl/artykuly/179610,instagram-hashtagi-z-glowa> [Accessed 20 September 2015].

Tifentale, A., 2014. *The Selfie: Making sense of the 'Masturbation of Self-Image' and the 'Virtual Mini-Me'*, [pdf] Selfiecity. Available at: <http://selfiecity.net/#theory> [Accessed 30 September 2015].

Troch, B., 2014. 'Co to jest hashtag (hasztag) i jak go używać w social media marketingu?' *Social Hero Blog,* [blog] 6 November. Available at: <http://www.social hero.pl/co-to-jest-hashtag-social-media-marketing> [Accessed 15 Septmeber 2015]

Tylikowska, A., 'Sezon narcyza sieciowego'. *Polityka,* [online] 8 October. Available at: <http://www.polityka.pl/jamyoni/1594491,1,samouwielbienie-internetowe. read> [Accessed 30 September 2015].

Weisma, A., 2015. 'Marc Jacobs is using Instagram to find potential models'. *Business Insider,* [online] 17 March. Available at: <http://www.businessinsider. com/marc-jacobs-instagram-casting-call-for-models-2015–3> [Accessed 28 September 2015].

Woolaston, V., 2015. 'Instagram is now bigger than Twitter: Photo sharing app has 400 millions users who share 80 million images every day'. *Daily Mail,* [online] 23 September. Available at: <http://www.dailymail.co.uk/sciencetech/ article-3245976/Instagram-bigger-Twitter-Photo-sharing-app-400-millions-users-share-80-million-images-day.html> [Accessed 30 September, 2015].

Wortham, J., 2013. 'My Selfie, Myself'. *New York Times,* [online] 19 October. Available at: <http://www.nytimes.com/2013/10/20/sunday-review/my-selfie-myself. html?_r=0> [Accessed 28 September 2015].

Zeberka.pl, 2015. 'Maffashion: real vs Instagram'. *Zeberka.pl,* [online] 28 February, Available at: <http://www.zeberka.pl/art/maffashion-real-vs-instagram-foto-34648> [Accessed 10 September 2015]

Mateusz Felczak

Presence and agency: Narratives of representation in modern computer role-playing games

This text focuses on computer role-playing games as a genre of electronic entertainment that highlights the player's presence in the digital environment. The article explores several issues tied to mechanical rules of interaction in video games. It investigates the connections between the software that governs and filters the images forming the player's field of vision. Secondly, it explores the interplay between the game's audio and visual elements. Then the article addresses the problems of racial and economic stigma in multiplayer online games. Finally, I argue that, in the modern AAA cRPGs, the representations of players are ultimately subjugated and evaluated by the game code.

Keywords: video games, avatars, agency, cRPG, interface

In this paper I will focus on issues which relate to the process of constructing an avatar and making it an extension of the player's agency, going beyond the simple equation of face with representation. I will not argue for removing our sense of vision from the analysis of video games, nor will I neglect the 'material perception' of the avatar. Quite the contrary: I will try to present evidence of players' extraordinary social investment in the visible aspect of avatars and the consequences this bears for the whole human/computer interaction process in the video-game medium. My main focus in this article will be on high-budget, 3D games, therefore I must take into consideration two roles of the modern triple-A RPG franchise game: the need to be recognised and the need to abide by the rules of persuasive interaction within the video game medium. My analysis will be based on a selected group of games available on the PC platform.

The mechanical rules of interaction

When speaking of how a game creates its internal rule system, one needs to define the meaning of game mechanics. For my purposes I will use Miguel Sicart's notion of game mechanics [see Sicart 2008], as a coherent system of responses to a player's actions based on the principles of the object-oriented programming paradigm. This definition enables us to perceive the player's digital embodiments not as finite objects that simply construct a visual representation, but as dynamic vessels, open to being re-situated and re-contextualised according to how the player frames the notions of identity and representation. Sicart's definition is also useful in discussing the player's agency. It calls into question the human impact on the code that operates the game engine and is responsible for the player's visualisation within the game world. There are video game genres that deliberately remove the avatar from player's field of vision [see Hitchens 2011] and interlock the two perspectives: the agent's and its distinctive 'place' in the game world. Such is the case with the FPS (First Person Perspective) genre, where the line of sight literally constructs the in-game world at the very moment it is being seen.

There are two aspects of this situation: one is entwined with the regime of perception and the work of interpretation, and the second is an example of Sicart's connection between mechanics and meaning.

In the first case, the game space must be interpreted as a field of signifiers, ready to be put to work by the player's act of exploration. The player's field of vision delineates options for interaction, guided by a simple mechanism: the player only has to look at a particular object for actions to become available, either through an on-screen indication (such as 'press <S> to collect the item') or by enabling a scripted event (as in cut-scenes triggered by the exploration process). It is important to note that, nowadays, the 'disembodied' or 'faceless' perspective of FPS games is often comprised of optional camera settings in modern cRPGs (like *Skyrim*) or TPP action games (*GTA V*). This enables the dynamic reassembly of players' point of view and their software-based perception toolkit during gameplay.

The second case derives from the main principle of video games' technical design: code efficiency and optimal hardware resource management. This is also disconnected from how a player's 'face' is represented in the game world, as it only factors in its presence in a particular area of the game. Put simply, the game world comes to life only when the algorithm decides that the player might see it. In some older games of various genres (an action-cRPG, *Diablo 2,* is one example) players can observe it by quickly utilising options that enable fast travel between the game areas, such as the teleport spell. During such swift exploration of the space it is easy to see groups of enemies suddenly changing from 'idle' to 'patrol' or 'hostile', as various characters catch the player's presence with noticeable delay. In such cases, it is not only the avatar's presence that shapes the whole game's dynamics, but, from the other perspective, the game 'recognises' the player's embodiment as an *axis mundi* which is the only incentive for the program to operate functionally.

This perspective seemingly puts the player in the centre of the interpretation process that constitutes the video game experience, but at the same time, it shows the limits of agency, which is always tied to the single perspective defined by the code. Even if the perspective can be switched, this effectively obfuscates the mechanisms which make the games coherent and playable. Therefore, before providing my definition of the avatar, I would like to investigate the formal and structural layers of a video game that make up the environment in which the metaphorical face of the player comes to be.

The voice of the avatar

Dovey and Kennedy [2006, p. 137] state that, when interacting with an avatar, the player is subject to a double embodiment in the game world. Therefore, game designers are confronted with the problem of translating a sensory experience into a game code, an artificial language that the player must intuitively recognise. In terms of the avatar's mechanics, the ability to speak is a feature that supports immersion in the game world, but it is sharply divided into two categories: scripted dialogues that are often read aloud after the player has chosen a particular line, and rule-based one-liners, exclamations or even onomatopoeias. The difference between these two types of voice lies in the mechanics of a game: for example, cries of pain or terror have to be scripted to match the actual danger the avatar faces in the game world. As such, encounters are often hard coded in the game and follow the simple if -> then paradigm. In the open world cRPG games in particular, this may lead to unintentional humour, which immediately disrupts the player's immersion in the story. Perhaps the most famous example of an awkward line of dialogue can be found in *The Elder Scrolls: Skyrim* [Bethesda 2013], which reads as follows: 'I used to be an adventurer like you, but then I took an arrow to the knee'. The designers decided to give almost every guard in the game the same story to share with the player – about a knee injury. That being said, without this voice the avatar would probably fall into the Uncanny Valley, a phenomenon in aesthetics that states that when an artificial being looks almost human (but not quite), it prompts strong negative feelings in the human observer. Voiceless characters are also frowned upon in the modding communities, where providing a set of custom-made sounds is considered indispensable to an immersive game experience.

It is important to note that the voice has many meanings; one of the more prevalent examples of the voice metaphor connects directly to agency and is used in the wider field of cultural and post-colonial studies. As we seem to see in game studies, avatars do not merely represent a player's body, but, due to the close ties with the game mechanics, they set boundaries for all of the actions that players can or cannot do. This suggests a more metaphorical meaning of the avatar's voice, or voice in general, which pertains to computation as an image of the social and political relations constructed in the community of consumers. According to T. L. Taylor, avatars

> facilitate not only the production of identities (shaping a persona through the look and actions of an avatar), but social relationships and communication as well. They are not neutral; indeed, their power lies in the very fact that they cannot be [Taylor 2009, p. 117].

As avatars are designed to give players the illusion of individuality, the urge to further enhance the avatar's social and persuasive role can be observed in areas that are very hard to reproduce on a satisfactory level, such as vocal communication. In computer games this is a highly complicated issue, ranging from in-game sounds to online player-to-player conversations. Nowadays, it is also an important part of the growing business of live streaming video platforms, such as Twitch.tv, where game commentary constitutes a substantial part of the experience. This aspect of avatars is seldom taken into consideration by game studies scholars, but I would argue that serves as a tentative link between the two main avatar roles outlined in the first part of this paper: recognition and interaction. Both have an ideological (i.e.: persuasive) role in positioning a constructed identity within the tastes of a particular community [see Bourdieu 1984]. An example in computer RPGs is the practice of voice acting during recorded play sessions (later archived in VOD format and/or broadcast live). This is different from the more traditional 'let's play' format, as it often aims to invent and read out the player's dialogue lines attributed to particular characters. At the same time, it hovers between immersion and a playful joke, especially in terms of cRPGs that enable control over multiple characters in a party, such as *Baldur's Gate.*[1] Yet another case of re-shaping the avatar according to one's particular needs, voice acting also connects computer RPGs to their social roots of traditional table-top RPG gaming. However, the collective aspects of how the avatar functions in game worlds introduce some rather problematic issues.

The face of the other

Another issue that needs addressing in terms of human representations in computer games concerns race and gender. In her study on players and avatars in *World of Warcraft,* Lisa Nakamura observes that, despite the apparent equality between various player figures in the game world, gamers in massively multiplayer online (MMO) environments find ways to stigmatise those who do not fit the

1 Cf. http://www.twitch.tv/unraveller/v/34073626 (accessed 10.01.2016).

average player type. As she writes in her text on ethnic and racial problems that are transferred through the medium of games:

> Player resentment against Asian player-workers results in a continual process of profiling other avatars to determine their status as legitimate leisure players or as unwanted farmers. […] Though players cannot detect other players' races by looking at their physical bodies, they constantly produce a taxonomy of behaviors that create new racializations of avatar bodies in digital space [Nakamura 2013, p. 192].

This insight into Asian players who play computer games not for leisure, but for profit (or as a penalty, serving "in-game time" instead of conventional jail time), offers a unique and often overlooked perspective on the avatar and its role in the communication process between game players. My point of interest is somewhat similar, though its scope is wider: I am interested in how the discourse of representation (in computer games and in other pop-culture media as well) is shaped and modelled by the paradigm of computation, which involves both hardware and software solutions that aim to promote a very specified mode of consumption. What our avatars can or cannot do is the result of a tailored ideological strategy which engages the player in something I will call the illusion of agency: repeated actions that create an illusion of progress, an illusion of identity and artificial subjectivity. Just as automatically repeating one action, like pressing a button on a camera, will not make anyone a better photographer, pushing an identical combination of buttons to fulfil the objective of a mission will not make our avatars any more proficient in these actions. This observation has already been made by such scholars as Jonas Linderoth [2013], but I would like to address a slightly different set of issues. My view is that very few games require users to actually improve their skill as players in order to improve the skills of the avatar. In other words, the transfer of the learning curve between the human agent and her digital embodiment is asymmetrical. The visual representation of the game code, manifested in the graphic user interface, creates an illusion of the player's ongoing progress by adjusting mathematical (in terms of the mechanics) and aesthetic (in terms of the avatar's appearance) data, which is only discursively connected to the changes in the game state. Game worlds are, from this perspective, a double illusion: that we, as players, improve our skills within their ruleset, and that our avatars develop and change as we progress. Therefore, our face in modern 3D computer games to some extent shares the same philosophical and sociological problems that are prevalent in modern, corporate-based societies: the alienation of work from the labourer and the disconnect between actions and their real outcome.

We can find some exceptions to the rule of this specific type of 'de-skilling' in titles like the *Dark Souls* series or some rogue-like games which implement a gradual curve of difficulty, but the prevalent pattern, especially in computer RPGs, is that the players are actually excluded from the process of meaning-making. In fact, there are whole computer game genres which have picked up on this problem as a core game mechanics issue: namely action RPGs, or, as some scholars prefer to call them, hack&slash games, notably the *Diablo* [Blizzard 1998–2015] franchise. In the third and latest instalment of the series, players can customise their avatar by altering the equipment's colour, shape or form. The important difference between actions performed by the avatars in these games and action in most other electronic entertainment products is that players create an appearance that they alone know is false: other players in the multi-player game environment only see the avatar's 'altered' façade. In the second edition of *Diablo* (2000), this situation was even more complex: by manipulating the pieces of equipment that gave bonuses to the avatar's statistics, one could effectively make certain pieces of gear invisible to other players. Though the visual feedback was still set by the code on the player's screen, when she joined the multi-player game, her avatar appeared to be 'naked' to the other game's participants. Such behaviour had an interesting impact on the player community – the 'bugged' appearance was quite coveted, both as a means of concealing particular equipment from other players in player vs. player games, and as a sign of being part of an 'elite' community that holds knowledge of this particular bug/glitch.

The false agency

If we think of the game as an apparatus, to borrow a concept from Norbert Wiener and Vilem Flusser [see Flusser 2011], we can see that the rules of communication impose certain technological and ideological affordances upon the persuasive rhetoric of video games. 'The camera (like all apparatuses that followed it) is computational thinking flowing into hardware' [Flusser 2011, p. 54]. The player's image in the game consists of bits of information, designed to convey messages to the other users and to the game world. If we apply Ian Bogost's procedural rhetoric [see Bogost 2007] to the analysis of players' avatars in RPG games, we discover that the medium's alleged persuasive power in high-budget RPGs is constrained by the ideological framework of both the software and the hardware. I would argue that it is also disconnected from the discourse of realism (or the relationship between the computer game and the 'real-world'), even though it often borrows themes, narratives or, in the case of the avatars, faces from the wide world of pop culture.

One of the more striking examples of this 'false presence' and 'false agency' in modern RPGs can be found in recently published high-budget cRPGs. In the *Dragon Age: Inquisition* [Electronic Arts, 2014], players have numerous options for shaping their playable character with very precise characteristics: from height and nose length, to eye colour and the position of cheekbones. The results of such avatar modifications of in-game playable characters often correspond with the prevalent pop-culture tropes in other media. Such is the case with the *DI:I* character, meticulously based on the appearance of Gwendoline Christie, the actress who plays Brienne of Tarth in HBO's *Game of Thrones* TV series. The *Dragon Age's* Brienne is an image from a narrative pasted into another narrative without consequences to the storyline or interaction between the player and the game world. From a game mechanics perspective, the avatar's aesthetics are meaningless here, turning the player's efforts into just another superfluous, expendable artefact. However, such playful alteration of the avatar may establish a gateway to a particular in-game performance involving the character, therefore – owing to the popularity of video games as such – becoming a significant contribution to the paratexts repository of the *Game of Thrones* franchise. I would argue that, in the case of each avatar that can be adjusted as the player wishes, there is a meaningful tension between the acting and playing. The notion of 'play' requires the game to be a coherent and nearly closed circuit of available actions and responses. As Mia Consalvo observes in her book,

> The rules of a videogame are contained within the game itself, in the game code. The game engine contains the rules that state what characters (and thus players) can and cannot do: they can go through certain doors, but not others; they can't walk through walls or step over a boulder (except maybe a special one); they can kill their enemies, but not their friends; and they must engage in certain activities to trigger advancement of the story and the game. All of these things are structured into code of the game itself, and thus the game embodies the rules, *is* the rules, that the player must confront [Consalvo 2007, p. 85].

In the aforementioned example of *Dragon Age: Inquisition* the *embodiment* of rules by the game is a very peculiar case of a man-machine interaction mediated by the avatars. I would argue that, from a purely mechanical perspective, the term 'avatar' can only be applied to characters that exert non-binary agency during the gameplay, i.e. those that can go beyond the friend-foe opposition. This conceptualisation is slightly different from the most prevalent definitions of the avatar, which stress the importance of a player's control over a particular game unit [see Kromand 2007, p. 400]. Consequently, my understanding of this term includes player's companions and recruitable characters, but excludes most of the enemies and placeholder individuals whose main purpose is to increase immersion in the game world. The rules, which are inextricably linked with the game code, are designed to coerce the player into certain ways of playing; especially in the cRPG genre, the definitions of failure and success are tied to the particular avatar's statistics (both mechanical and aesthetic). Even in older games of this genre, where visual representation was limited to a handful of animated pixels, one of the statistics usually corresponded to both physical attractiveness and persuasive power. An illustration of this can be found in the *Dungeons & Dragons*-based Infinity Engine games, such as *Baldur's Gate* and *Icewind Dale*, where the 'charisma' stat was responsible for determining initial responses in interactions with other characters. Such a strong discursive connection between in-game performance and the statistics of the avatar still holds sway in the modern cRPGs. Some of them, including the latest iterations of the *Dragon Age* series, even emphasise the avatar-to-avatar interactions of the characters with assigned roles in the story, making them the main part of the gameplay, as opposed to the common design practices of this traditionally combat-focused genre. In theory, this should transfer the main incentive to play from engaging in statistics-driven combat to enacting the character controlled by the player. The former form of interaction, however, often involves scripted events and cut-scenes, where the computer, not the player, guides the avatars' actions. The main obstacle to the player's agency *vis-a-vis* the avatars lies in the code, which is ultimately designed to govern all the important aspects of the game: the story, mechanics and various objects, along with their embodiments in the game space.

The rhetoric of efficiency

There is yet another important aspect of the avatar's impingement upon the game's virtual environment: the time frame in which the production of meaning takes place. The graphic model which depicts and represents a particular being within the game's diegesis is ultimately judged not during its actions and performance, but after it is already outside of the player's scope. For example, from the game's perspective the NPCs (non-playable characters) in modern cRPGs are evaluated according to their usefulness in particular situations. Only after the players see the outcome of their behaviour are they able to evaluate them. This process can be summarised as forming a teleological interpretation which corresponds to game theories that emphasise the game's performance over its other aspects.

I would argue that, in the cRPG genre, the representations of players are subjugated to a process of negotiating meaning within the game's narrative. In other words, the fact that our in-game avatars connote particular meanings is a result of their role in the game diegesis, which is always determined by a pre-formatted set of rules. The players can play against these rules, but ultimately the performance of their avatars is to be judged and evaluated by the game code. This fact coincides with Ian Bogost's famous notion of procedural rhetoric.

> Procedural rhetoric is a subdomain of procedural authorship; its arguments are made not through the construction of words and images, but through the authorship of rules and behavior, the construction of dynamic models. In computation, those rules are authored in code, through the practice of programming [Bogost 2010, p. 29].

In Bogost's analysis, rhetorically significant mechanics do not have to be implemented with the intention of a particular outcome, much as the narratives of representation imposed on the players do not have to present a coherent ideological content. There is, however, a governing instance that permeates the rules of the computational rhetoric that presides over players' in-game representation, and I would argue it is connected with narratology, i.e. it derives directly from the affordances defined by the story. Because the avatars in cRPGS are always presented in a broader context – even when it comes to player-designed characters we are often provided with their history and affinities within the fictional in-game environment – the process of internalising their visual aspects and attributes highly depends on the interpretation of a particular narrative. The narrative context and the avatars are therefore entwined in a rhetorical set of assets that represents the game's agenda. In *Skyrim*, players have plenty of say in the physical appearance of the avatar and her skills and attributes, but are forced to acknowledge 'their' paranormal ancestry and seemingly important role in an epic narrative. From a marketing perspective, this combination provides great incentive for various paratexts from the community, which greatly contributes to the media buzz around the title. *Skyrim*'s story about the Dragonborn or the *Fallout* or *Baldur's Gate* series narratives about being the chosen one or the child of a god seem to be mere reiterations of a well-known heroic *topos*, yet they provide a common platform for players to compare their avatar's contributions and their role in the particular game worlds.

Agency, presence, control

After taking into consideration the features of the player's representation in the game world and the narrative context they are placed within, we should investigate the possible correlation between the narrative and mechanical components of the game which derive their meaning from the intrinsic rhetoric. However, to analyse this issue effectively we need a theoretical framework that will allow us to interpret both the surface and the hidden mechanisms enabling the production of meaning in a video game. One such approach, albeit rather radical in its presumptions, is Kristine Jørgensen's notion of game world interfaces. She argues that

> the game world has many properties. It is the interface to the game system, and for this reason it is representational of it. At the same time, game worlds are word constructs suggesting ecological environments. Game worlds are also games and for this reason are built around the logics of games [Jørgensen 2013, p. 55].

Such a holistic approach allows us to apprehend the game interface as an extensive platform for meaningful interaction between game agents, among which two are of crucial importance in single-player cRPGS: the human player and the code-driven machine. To link these two we need to develop a theory that covers numerous layers of the game without favouring any of them. I would argue that such a role can be filled by operational logics, introduced by Noah Wardrip-Fruin. It can be summed up as a method of tracing and interpreting patterns of interplay between data, process, surface, interaction, author and audience [see Wardrip-Fruin 2009, p. 13]. It is crucial to note that the patterns between computational elements (for example, different pieces of data) are given no preference over the human-computer interaction. The only visible outcome of the data processing is on the surface, in the various displays and tangible hardware. The most important thing about this model is that it broadens our perspective in analysing the avatar's agency and breaks with the prevalent model of linear interpretation based on the human input and computer output circuit. In the cRPG avatars, the 'author' is responsible for programming visuals and data, but it is set in motion as a direct result of a real-time negotiation between human and computer agents. This software studies perspective may provide an explanation for Jørgensen's observation that 'during our first encounter with a new game, the game world might appear both alien and familiar at the same time' [Jørgensen 2013, p. 78]. It is familiar because it is comprised of elements which, on a primary, conceptual level, were designed by a human (author), but it is alien – and the avatars are also alien to the players – because ultimately they are governed by a set of rules established on a computational level, as an interpretation (or conceptualisation) of data through processing. In other words: each computer game (and, to some extent, every interactive program as well) has a human actor on both ends, but in single-player cRPGs meaning is conveyed in the absence of the initial human author. Consequently, the player guiding an avatar and, for example, asking a merchant to buy goods in *Skyrim,* conducts a stimulated interaction between the two sets of computational objects. What is more, both of them create a double illusion of complete agency and visuality in order to hide the fact that the interaction is subjugated to the predefined rules of binary logics.

Conclusion

It the end, we arrive back at the basic issue concerning the game's interface: what appears to be right before our eyes, on the visible and transparent space of the screen, in fact lies somewhere between the player's affordances and the machine's presets. If, then, as Alexander Galloway [2006] opines, 'virtual worlds are always in some basic way the expression of utopian desire', the desire to be recognised in the form of an avatar must be among the hardest to achieve. Navigating between the software and the hardware restrictions, as well as the ideological narratives supported by the obtrusive game mechanics, requires great caution and self-awareness to achieve an ostensibly simple goal – inserting our face in the games we play.

Bibliography

1. Bogost, I., 2010. *Persuasive Games. The Expressive Power of Videogames*. Cambridge, Mass.: MIT Press.
2. Bourdieu, P., 1984. *Distinction: A Social Critique of the Judgement of Taste*. Cambridge, Mass.: Harvard University Press.
3. Dovey, J., Kennedy, H., 2006. *Game Cultures: Computer Games as New Media*. Maidenhead, Berkshire, England: Open University Press.
4. Flusser, V., 2011. *Into the Universe of Technical Images*. Minneapolis: University of Minnesota Press.
5. Galloway, A., 2006. *Gaming. Essays on Algorithmic Culture*. Minneapolis: University of Minnesota Press.
6. Hitchens, M., 2011. 'A Survey of First-person Shooters and their Avatars', *Game Studies*, Vol. 11, Issue 3. Available at: http://gamestudies.org/1103/articles/michael_hitchens.
7. Jørgensen, K., 2013. *Gameworld Interfaces*, Cambridge, Mass.: MIT Press.
8. Kromand, D., 2007. 'Avatar Categorization', Situated Play. Proceedings of DiGRA 2007 Conference. Available at: http://www.digra.org/wp-content/uploads/digital-library/07311.16435.pdf, pp. 400–406.
9. Linderoth, J., 2013. 'Superheroes, Greek gods and sport stars: Ecological empowerment as a ludo-narratological construct'. In Mitgutsch, K. Huber, S., Wimmer, J., Wagner, H. G., & Rosenstingl, H. (Eds.) Context Matters! Proceedings of the Vienna Games Conference 2013: Exploring and Reframing Games and Play in Context. Vienna: New academic press.
10. Nakamura, L., 2013. 'Don't Hate The Player, Hate The Game: The Racialization of Labor in World of Warcraft.' in *Digital Labor: The Internet as Playground and Factory*, ed. Scholz T, New York: Routledge, pp. 187–204.

11. Sicart, S., 2008. 'Defining Game Mechanics', *Game Studies,* Vol. 8, Issue 2. Available from: http://gamestudies.org/0802/articles/sicart.
12. Taylor, T. L., 2009. *Play Between Worlds. Exploring Online Game Culture.* Cambridge, Mass.: MIT Press.
13. Wardrip-Frun, N., 2009. *Expressive Processing. Digital Fictions, Computer Games, and Software Studies.* Cambridge, Mass.: MIT Press.

Polish Studies – Transdisciplinary Perspectives

Edited by Krzysztof Zajas and Jarosław Fazan

Vol. 1 Artur Płaczkiewicz: Miron Białoszewski: Radical Quest beyond Dualisms. 2012.

Vol. 2 Kinga Kosmala: Ryszard Kapuściński: Reportage and Ethics or Fading Tyranny of the Narrative. 2012.

Vol. 3 Michał Nowosielski: Polish Organisations in Germany. Their Present Status and Needs. 2012.

Vol. 4 Krzysztof Zajas: Absent Culture. The Case of Polish Livonia. 2013.

Vol. 5 Magdalena Sitarz: Literature as a Medium for Memory. The Universe of Sholem Asch's Novels. 2013.

Vol. 6 Barbara Przybyszewska-Jarmińska / Lech Sokół (eds.): Poland and Artistic Culture of Western Europe. 14th–20th Century. 2014.

Vol. 7 Katarzyna Fazan / Anna Róża Burzyńska / Marta Bryś (eds.): Tadeusz Kantor Today. Meta-morphoses of Death, Memory and Presence. Translated by Anda MacBride. 2014.

Vol. 8 Andrzej Hejmej: Music in Literature. Perspectives of Interdisciplinary Comparative Literature. Translated by Lindsay Davidson. 2014.

Vol. 9 Grzegorz Niziołek: Warlikowski: Extra Ecclesiam. Translated by Soren Gauger. 2015.

Vol. 10 Ryszard Koziołek: Sienkiewicz´s Bodies. Studies of Gender and Violence. Translated by David Malcolm. 2015.

Vol. 11 Wojciech Tygielski: Italians in Early Modern Poland. The Lost Opportunity for Modernization? Translated by Katarzyna Popowicz. 2015.

Vol. 12 Dariusz Jarosz / Maria Pasztor: Polish-French Relations, 1944-1989. Translated by Alex Shannon. 2015.

Vol. 13 Urszula Augustyniak: History of the Polish –Lithuanian Commonwealth. State – Society – Culture. 2015.

Vol. 14 Piotr Sobolczyk: Polish Queer Modernism. 2015.

Vol. 15 Jacek Soszyński / Agnieszka Chamera-Nowak (eds.): Book versus Power. Studies in the Relations between Politics and Culture in Polish History. Editorial assistance by Dan Embree. Translated by Jacek Soszyński 2015.

Vol. 16 Wojciech Kriegseisen: Between State and Church. Confessional Relations from Reformation to Enlightenment: Poland – Lithuania – Germany – Netherlands. Translated by Bartosz Wójcik and copy-edited by Alex Shannon. 2016.

Vol. 17 Urszula Sowina: Water, Towns and People. Polish Lands against a European Background until the Mid-16th Century. Translated by Justyna Woldańska. 2016.

Vol. 18 Grzegorz Krzywiec: Chauvinism, Polish Style. The Case of Roman Dmowski (Beginnings: 1886-1905). Translated by Jarosław Garliński. 2016.

Vol. 19 Andrzej Sakson: Von Memel bis Allenstein. Die heutigen Bewohner des ehemaligen Ostpreußens: Memelland, Kaliningrader Gebiet, Ermland und Masuren. Übersetzt von Marek Drewnowski. 2016.

Vol. 20 Antoni Mączak: Unequal Friendship. The Patron-Client Relationship in Historical Perspective. Translated by Alex Shannon. 2017.

Vol. 21 Olga Szmidt / Katarzyna Trzeciak (eds.): Face in Trouble – From Physiognomics to Facebook. Copy-edited by Soren Gauger. 2017.

www.peterlang.com

www.ingramcontent.com/pod-product-compliance
Ingram Content Group UK Ltd.
Pitfield, Milton Keynes, MK11 3LW, UK
UKHW041829100926
8556IPUK00002B/87

9 783631 676233